KNIGHTS OF THE BROADAX

KNIGHTS OF THE BROADAX

The Story of the Wyoming Tie Hacks

by

JOAN TREGO PINKERTON

The Caxton Printers, Ltd.
Caldwell, Idaho 83605
1981

Pinkerton, Joan Trego.
Knights of the Broadax.

1. Tie hacks—Wyoming—History. I. Title.
HD8039.L92U573 338.7′6742 79-57239
ISBN 0-87004-283-1 (pbk.)

Lithographed and Bound in the United States of America
by
The Caxton Printers, Ltd.
Caldwell, Idaho 83605
135990

In memory of my father, A. B. Trego

CONTENTS

ILLUSTRATIONS

ACKNOWLEDGMENTS

This book was a long time aborning. It began when I realized there were few alive who even remembered the Tie Hacks, and I felt a need to preserve the memories for coming generations.

The majority of the pictures were taken by my father, who called himself just an amateur photographer but who captured the flavor of life on the mountain and the grandeur of the surroundings as well as anyone could. A meticulous, well-organized man, he left behind a legacy of negatives and prints that bring to life the story of the Tie Hacks.

Even so, without the gentle prodding of dear friend Suzanne Brockmeier, *Knights* would still be that book I was going to write someday.

A patient and encouraging husband helped bring it to fruition. Dick spent many hours alongside me in the dark-room, enlarging and developing, and was chief copy editor.

Encouragement and assistance came from many friends and especially from my mother, who helped identify people and places. I am deeply grateful to those who loaned old pictures and who, over a cup of coffee, recalled "the way it was." It was glorious.

KNIGHTS OF THE BROADAX

Chapter One

A LEGEND IN THEIR TIME

Their broadaxes skillfully hewing trees into railroad ties, the Tie Hacks of Wyoming cut a swath through more than just the timber of the high mountains. These rugged woodsmen carved their rightful place alongside other American folk heroes with their legendary craftsmanship and joyous boozing.

They were lumberjacks of the lodgepole pine, who plied their trade with brawn, crosscut saw, and ax, long before the days of the power saw.

What set a Tie Hack apart from the traditional lumberjack was what he did with the tree after he sawed it down. His special talents with the ax began at that point. First he limbed the tree, then scored the sides, standing atop the fallen pine and swinging his double-bitted ax with unerring aim beside him the length of the tree. Two sides were scored with a series of parallel slashes several inches apart. These cuts smoothed the way for the final step: the hewing of the tie to a polished smoothness with another formidable tool, the razor-sharp twelve-inch-wide broadax which was the mark of the Hack's trade.

Unerring aim and feet pointed straight ahead were *needed* — the heavy axes could as easily catch a stray toe as a chip of wood. So accurate were the men at their craft the ties gave the appearance of having been planed, and they were invariably the right size, though no Hack deigned to measure his product.

The railroad inspectors expected the depth of the tie to measure at least seven inches, and no more than seven and three-fourths inches, to fit evenly into the bed of the railroad.

Winter months were devoted to tie-making. "Squeak Axel" Johnson was typical of the Hacks, who felled their own trees and carved them into precise eight-foot-long ties.

The process began with limbing and scoring the felled tree with a double-bitted ax. Ted Berg pauses after scoring one side. Behind him is the stack of ties cut so far that day.

The other two sides of the tie were left rounded, peeled of bark. From two to five eight-foot ties could be cut from one tall lodgepole pine.

Back in the early 1900s a good Hack worth his salt could cut fifty ties a day, at the going rate of ten cents per tie. The price rose through the years, until World War II demand for railroad ties pushed the price up to fifty cents apiece!

There were tie-making operations scattered throughout Wyoming. The one at DuNoir, high in the remote Wind River Mountains near the Continental Divide, was run by the Wyoming Tie and Timber Co. From the early 1900s into the 1940s this company produced over ten million ties, mostly hand-hewn, for the Chicago and Northwestern Railroad.

The Hacks employed there were mostly Swedes and

Smoothing the surface of the tie was accomplished with the broadax. John Berglund swings his seven-pound ax deep into the wood. After the ties are hewn from this tree, the bark will be peeled from the two rounded sides. The ties were as smooth as if they had been planed by the time the Hacks, real craftsmen, finished them.

Norwegians who had come to America around the turn of the century. No assembly-line process this; they not only carved the ties themselves, but each summer they herded the winter's cut down the swift waters of the Wind River in a spectacular 100-mile river drive to the tie yards in Riverton.

These strong men, who did hard, back-breaking work, often in lonely solitude, produced other results not quite so measurable but infinitely more colorful, as recalled in the Beowulfian tales of their lusty brawling, hard drinking, and feats of strength.

This is the story of the DuNoir Tie Hacks and their river drives as told to me by the handful of persons still living who worked in the tie camps, and as I remember those days myself, because I was privileged to be a child growing up in their midst.

Photo courtesy Lydia Olson

Woods Boss Martin Olson *(right)* and William H. ("Billy Mac") McLaughlin, assistant superintendent of the Wyoming T&T, are dwarfed by the surrounding piles of hand-hewn ties — over 92,000 of them — in this photo taken in the early 1920s in the Riverton yards.

From high atop the cliffs on Lookout Mountain to the east of Headquarters, the cluster of cabins and buildings that made up DuNoir is dwarfed by the surrounding mountains of the Wind River Range. In the background are the the snow-covered peaks of the Absaroka Range. The road leading out of the picture to the right was the only avenue off the mountain, winding five miles down to the highway. Meandering up to the left, the road led to some of the upper camps. The picture was taken in 1945. The author, Joan Trego Pinkerton, sits at lower left, twins Jean and Ann White above her.

Stark, windswept early winter view of the Wind River Valley. Photograph taken from top of Warm Springs Mountain just before reaching Headquarters.

I witnessed the passing of an era in America's history.

I was in the third grade when my family — mother, father, and myself — moved to DuNoir, which was headquarters for the Wyoming T&T. My father, A. B. Trego, was to be accountant and secretary for the firm. It was 1936 and we were lucky he had the job, but it was like moving to another world for the three of us, coming from New York.

We left the highway ten miles west of Dubois and turned onto a narrow, rutted dirt road that wound around an agonizing five miles of switchbacks up Warm Springs Mountain to get to our new home. The midafternoon sun filtering through the golden-dollar aspen leaves dappled the road ahead of us. Finally we reached the crest of the mountain; another mile and we had dropped over the top to Headquarters.

Headquarters — main camp complete with store, office, post office, and schoolhouse — housed about a dozen families. It was a collection of hand-hewn log cabins spread in neat rows across some eighty acres atop a wind-swept mountain plateau. Devoid of trees or vegetation other than sagebrush, it was nevertheless surrounded by the most magnificent scenery. There were mountains rising on every side, covered with heavy stands of lodgepole pine, that tree so uniquely suited for the manufacture of the railroad tie.

Warm Springs Creek snaked along the foot of Union Pass Mountain, just opposite Headquarters. Union Pass had been traveled by early pioneers moving west to Oregon. In later years I often skied its steep inclines and never ceased to marvel at the courage and persistence of those people who labored to cross the continent with their wagons and ox-teams.

We arrived in the late afternoon, just as the little, one-room schoolhouse was emptying for the day. I remember the curious stares of the youngsters who were to become my classmates as our car drove by the school.

That first night we were to be the guests of the Sorn Pederson family for dinner. Sorn was the blacksmith for the company, keeping not only the machinery in repair but the some forty workhorses in horseshoes.

The Pedersons had one son, little Joey, whom I remember only for his perspicacity when he was a first-grader. The teacher, a young enthusiastic woman eager to impart learning and culture to her charges, was trying to teach the word "flower" to the youngest in her care.

"You know, Joey," she coaxed, "what your mother puts in the center of the dinner table." Young Joey eyed the word on the blackboard doubtfully and finally blurted, "I dunno, but that sure as hell don't look like ketchup to me."

The cabin we moved into was exactly like all the others: four rooms consisting of two bedrooms, a living room, and a kitchen, each opening into the other with no connecting hallways. Built of chinked logs on dirt-and-wood foundations, the interior walls were covered with beaverboard. Battleship linoleum — brown — covered the floors. The houses had electricity, telephones, and cold running water. Only the Ahlbergs, who ran the laundry for the men, had an indoor flush toilet.

And the VanMetres, of course. VanMetre was president of the company. The family spent only the summers in DuNoir, going back to their home in Illinois every fall because the main offices of Wyoming T&T were in Chicago.

Each house had its own privy set back from the house. (Ours was a three-holer in graduated sizes, which always brought visions of the Goldilocks story about one for Papa Bear, one for Mama Bear, and one for Wee Baby Bear.) We took baths twice a week by heating water in an old boiler on the wood stove and pouring it into a rubber, collapsible bathtub set up in the center of the warmest room in the house — the kitchen. The tub was luxuriously longer than today's porcelain counterpart. It could be emptied only by dipping all the water into the kitchen sink. Then it was folded up and

VanMetre summer home, owned by the president of the Wyoming Tie & Timber Co. This was the only cabin in Headquarters with indoor plumbing, other than the laundry.

The Lone Pine. Twisted by the elements, this sentinel greeted travelers arriving at the top of the hill after driving up the Tie Camp road.

put away until the next bath night. It was, actually, an oversized bathinette such as used by modern mothers for their babies.

The company provided the housing for $14 a month, including electricity, ice twice a week, garbage removal, and all the wood needed for heating and cooking for the year. The wood was chopped and stacked by one of the employees in a neat woodpile next to each outhouse. My daily job, like all the other children in camp, I suppose, was to keep the family woodbox full.

Looking out our windows we could see other structures that made up Headquarters — the blacksmith shop, several hay barns, and storehouses. There was a small, forbidding-looking log hut built back into the side of a hill and topped with sod that I was warned to never go near — it housed the dynamite used for blasting new roads and breaking up tie jams in the river.

Beyond the storehouses the plateau dipped down to the creek, and it was along the creek the barns and corrals were built, as well as the sprawling log houses where the Ahlbergs maintained a laundry.

The Tie Hacks didn't live at Headquarters. These cabins were for the families of the management, the teacher, a few of the sawyers, the blacksmith — those whose work kept them in Headquarters. The men themselves lived in outlying camps near the site of timber they were cutting at the time. Each of the camps had a cookhouse, bunkhouse, and horse barn. The few married men built their own cabins for their families.

The bunkhouses were furnished in bleak simplicity — with bunks, that's all. No chairs, no tables. Great readers of

Late spring at Headquarters found the low areas still filled with snowbanks. Note the work-horses grazing around the cabins.

the Scandinavian newspapers and magazines received in the mail, the men lay on their bunks to read, sat on them to play cards. They weren't the only ones inhabiting the bunks. There were bedbugs galore. One Hack was heard to grumble, "I don't mind the bedbugs biting me, but I sure hate their cold feet walking all over me."

Some of the men preferred to batch rather than live in the bunkhouse and eat in the cookhouse. Disdaining this structured society, they built their own stark, mean cabins away from the camps, often by as much as half a mile. Here they lived in nearly complete isolation.

Cabin fever was not unknown during the long winter months, particularly among the bachelors. Two old curmudgeons sharing a cabin finally reached the limit of their toleration for each other. A blizzard raged outside; no one could get to work. Ole took a knife and scratched a line

Winter view of Headquarters camp, west end, taken from Union Pass Trail, the route taken by early pioneers in their trek to Oregon country. The Absaroka Range is in the background.

A temporary summer camp, erected at the site of a new cut

down the exact middle of the cabin, haughtily proclaiming his side inviolate. Angrily nodding his head, Andy agreed. Each would do his own cooking, light his own lamp, keep his own counsel. But with only one stove they had to take turns building and stoking fires. Ole even braved the storm to fetch water dipper by dipper rather than fill a bucket that Andy might share. Thus began their peculiar life of self-imposed silence with each other which continued long after the storm abated.

Headquarters was almost awash in creature comforts compared with these outlying camps. For the Tie Hacks, life in the backwoods without electricity, running water, or telephones may have been hard, but it was their way of life and they accepted it stoically. For the few wives trying to create

Three cabins at one of the outlying camps nestle under the trees in the lee of a hill

homes and raise families, the drudgery of the daily routine must have been overwhelming at times.

The simple names assigned to those clusters of cabins indicated which small creek ran through that camp. There were South Fork, Trapper's Creek, Spring Creek, Crooked Creek, Wild Cat, and myriad others over the years, all constructed for use only while the timber was being cut in that area. Water had to be hauled in buckets from the creek for cooking, washing, drinking. Candles and kerosene lanterns supplied the only light after dark. The wood cookstove provided the only heat in the cabins, few of which were more than one or two small rooms. Rough wood planks formed the walls, ceilings, and floors. The cabins housing families were easily identifiable by that traditional woman's touch: curtains at the window and a pot of red geraniums on the sill.

Wild Cat Camp, like all the others, often could only be reached on skis in the heart of winter.

On opening day in 1938 the little, one-room log schoolhouse at Headquarters was overflowing. There were more than twenty students in all eight grades (and baby brother, front row) for the harried teacher to cope with. The only athletic equipment was the swing set in the background.

If there was a child old enough for school, arrangements were sometimes made for the youngster to board for the winter with a family living at Headquarters. Or a youngster might just stay out of school that year. Occasionally the whole family moved to Headquarters for the winter.

One September day school opened with nearly twenty students, ranging in age from five to seventeen, sitting expectantly at their desks. This would be an exciting year, with so many for one teacher to handle. It was too exciting for the teacher — we went through four different ones that year. Enrollment never before or after reached that peak. Ordinarily there were fewer than ten in all eight grades.

When Headquarters was first constructed in 1928 there

In the winter the hill beside the school was great for sledding and skiing. The school outhouse had two sides — one for the boys (door open) and one for the girls.

It will be several months before this old Model T will be thawed out enough to drive

had been a need for a big cookhouse and several bunkhouses. After a few years these buildings were no longer used on a regular basis. At Christmastime the cookhouse became the theatre for the school's annual production of Dicken's *Christmas Carol,* followed by the long-awaited appearance of Santa Claus to distribute the mysterious packages under the tree to all the children of company employees.

Those magnificent Christmas boxes! I can see them still! To my child's eye they were immense. Earlier in the fall the president of the company, VanMetre, would have shopped in Chicago for the gifts, then mailed them to the woods boss for his wife to decide what would fit whom. There might be

Old ranch gate guards entry to the upper DuNoir River area. Wyoming Tie and Timber Co. moved its Headquarters camp from the base of mountains in the background to the top of Warm Springs Mountain in 1928.

snow jackets, always new mittens, and a giant box of crayons. Maybe dolls for the girls, trucks for the boys. There were books and candy. One year there was a sled for each child!

A few nights after the children's party, the cookhouse became a dance hall for the yearly bacchanalian event for all the Tie Hacks. Looked forward to for months, this all-night revel drew visitors from miles around. They would come by team and sled if Tie Camp was already snowed in. Music was usually provided by those Hacks like Richard Eastlund who played accordion. Eastlund was usually so inebriated by the time the sun was coming up (the signal the dance was over) he'd be lying on his back, still playing the accordion draped across his stomach.

"Punch" for the party was strong stuff consisting only of whiskey and a couple of lemons and oranges. It was stored in ten-gallon water cans and kept next door in the bunkhouse until the bowl needed replenishing.

On one such night of revelry, the man sent to carry over another can of punch found Brady Kjelmo had already made severe inroads into the liquid. He had pulled the can over next to his bunk, where he lay happily dipping into it, getting drunker by the cupful.

Chapter Two

WINTER ON THE MOUNTAIN

Long, harsh winters brought families in Headquarters together with skiing, card playing, and for the women the inevitable sewing club. The early years on the mountain found the camps often snowed in for up to three months, with no way of getting out other than by team and sled, or by skiing. Skis were the most efficient means of moving about and also provided recreation for children and adults alike. The kids

The simple log cabins at Headquarters housing the employees of the company stayed snug and warm even through the most vicious storms, heated by wood stoves of the kind now facing a popular revival in the energy-conscious age.

Winter in the high mountains found the Wyoming T&T Headquarters camp snowbound as much as three months out of the year in the early days. The company's only gas pump stands useless, surrounded by snow. The mountain rising in the background is Union Pass, one of the alternate routes used by immigrants trekking westward over the Oregon Trail.

skied to school, went out to ski at recess, skied home after school, then joined each other to ski until suppertime. Occasionally there were even romantic moonlight skiing parties.

A 1934 Riverton weekly newspaper, *The Riverton Review*, noted under a column titled "Tie Camp Topics" that "several women had an impromptu ski party on a local hill and tea was served at the Wester home."

In December 1937 that same newspaper noted, "Spring Creek, South Fork, Trapper Creek, and all the other small communities above Headquarters have been snowed in since the last storm. From now until spring the residents will have no way of leaving their homes other than by skis or using horse-drawn sleds. There is considerable rueful dismay because the snow came so unusually early this year."

Winter's mantle of white meant isolation but created unparalleled beauty. A bright moonlit night might find the men

When snow covered the roof of this high mountain cabin it was time to dig a tunnel entrance to the door.

After every new storm, new paths had to be dug to the woodpile (*front*) and to the outhouse (*center, nearly hidden by snow*). Hanging clothes on the nearly unrecognizable clothesline (*right*) was going to be a problem, too.

and the few families hitching a team and sled and traveling to an adjoining camp for an impromptu dance in the cookhouse. Snug and warm under blankets, sled runners cutting crisply through the glistening snow, and the bells on the horses' harnesses repeating the rhythm of their patient plodding, the revelers would entertain each other with song and stories of the Old Country. Someone in the camps could always be counted on to play the accordion for the schottisches and hambos the men danced with such gusto — danced with each other if no woman partner was available.

Since the ties were cut during the winter, it was easier for the men to negotiate the deep drifts on skis. Most of them had learned to ski in the Old Country and were as much at home on the long wooden boards as they were on their feet.

"Boards" is the apt description for those early-day skis.

In this picture taken from Union Pass, the west end of Headquarters glistens under its mantle of winter white. The U.S. Ranger Station is in the forefront, surrounded by its own fence. Behind the enclosure and to the left is the one-room log schoolhouse where the Wyoming T&T employees sent their children to learn the three Rs.

Most of them were homemade and bore little resemblance to today's sleek, narrow cross-country counterparts. The skis were wider and much longer and of course were made entirely of wood, since plastic hadn't yet been invented. Various methods were employed to assist the skier in moving uphill, such as animal skins strapped on the bottom of the boards so the fur gripped the snow while climbing.

The men seldom used a set of poles. Usually they had just one very long pole used for balance and to act as a brake on steep slopes. When they wanted to slow down, they just straddled the pole and sat back.

By 1946 a small portable ski lift set up near Headquarters had become a popular weekend attraction for skiers in the Wind River country.

One man in particular was poetry in motion on skis. Syver Gottenburg could soar like a bird off any jump, maneuver any hill with grace. Skiing for most of the men was a necessity — a way to get to work. But for Syver it was a skill in which he gloried. He knew the secrets of waxing and could walk straight uphill on his skis. At the top he would stamp his skis, rub them back and forth a few times, then sail downhill as though on wings.

Fortunately, there were few serious accidents. But one winter day a slow procession of men pushing and pulling a sled made its way to the company store. One of the Tie Hacks had been caught under a falling tree, and it had crushed his chest. The accident had happened some eight miles back in the woods, and his fellow workers had loaded

When the snows started, out came the favored method of transportation to work, to school, anywhere — long, wide skis, far different from today's sleek counterparts. "Fancy" bindings were of the bear-trap variety such as worn by Syver Gottenberg, above. More common were simple lash-ups made from a one-inch piece of rubber innertube slung around the heel and anchored over the leather toe strap. Syver, an accomplished skier, often spent his Sundays soaring through the air over jumps he had painstakingly built himself.

Syver packs his own ski slope

him onto a bobsled and pushed it all the way to Headquarters by hand. Once there they carried him carefully from the sled into the middle of the store, where they laid him on a makeshift bed on the floor.

While phone calls were made to summon an ambulance, his worried friends gathered around him, sat him up, and poured whiskey down his throat in an attempt to cheer him up. The man survived not only the injury but the treatment. He had to be taken by team and sled to the foot of the mountain five miles away, where the ambulance met the sled. From there it was still a long hundred miles to the nearest doctor at Riverton or Lander.

Because of that distance, people just didn't get sick very often, and only severe injuries required treatment beyond what the Red Cross emergency kit and ingenuity could provide.

It was an important day when the company invested in a large Allis-Chalmers bulldozer that could be used to plow the road from Headquarters to the highway in the winter. It was also used in the woods for road building, and in the spring the big cat speeded up the process of breaking out landings for the river drive.

Even if the roads weren't closed by snow, spring thaws and fall rains made them impossible quagmires during the daytime. They could be traveled then only at night or very early morning while the ground was frozen.

One rainy day a Hack misjudged and sliced two toes off his foot along with the limb he was cutting from a tree. He, too, was brought into the store, where he sat all day, suffering in typical Scandinavian silence, waiting for night, when

Situated close to the Continental Divide, Headquarters and the outlying camps were at an elevation of over 8,000 feet. Winter's fierce winds whipped the snow into grotesquely towering snowbanks which grew ever taller as the winter progressed.

The purchase of an Allis-Chalmers bulldozer by the company in the early '40s brought a significant change to the lives of the people who lived at Headquarters — the road off the mountain could be kept plowed most of the winter, ending the usual months of isolation.

the temperature would drop below freezing and the road would become firm enough that he could be taken off the mountain. Waiting for him at the highway would be genial "Billy Mac," William McLaughlin, superintendent of the company's tie yards in Riverton, a hundred miles away. Billy Mac was often pressed into duty as chauffeur when emergencies arose.

For the Hacks there were two standard patent medicines used for any ailment. If the problem was internal the Hack swallowed a dose of Curaco, and if it was external he rubbed on Oleoid.

Oleoid was a foul-smelling liniment probably first perfected for horses but found to be equally effective on human aching muscles.

Curaco was considered an all-purpose tonic taken for

After a fresh snowfall, moose have wandered along the base of this snowdrift in search of feed.

stomachaches, as a laxative, or to calm upset nerves. Upon reflection, it was probably Curaco's high alcohol content (14 percent) that calmed the nerves. Sometimes coming down off a binge, if liquor wasn't available, a drink of Curaco would help a Hack sober up.

One day a celebrating old-timer complained to his friend Helmer that someone had been in his cabin and had drunk all his Curaco. The next day the now-sober Hack sheepishly admitted to Helmer that he guessed he must have drunk it all himself because he had spent the biggest part of the night "in the little house" — his euphemism for outhouse.

Toothaches were to be endured as long as possible. Tales of a highly unorthodox form of layman dentistry were told about the early days in the tie camps. If the pain became too

A camp robber snatches a crust left him by a thoughtful human

much to bear, the Hack would anesthetize himself with whiskey and persuade a sympathetic friend to pull the offending tooth. After a few drinks to bolster his courage, the Hack-turned-dentist usually had difficulty identifying just which tooth was the culprit. To be on the safe side, he might pull one or two on each side of the one under suspicion — just in case.

Many of the men ended up with few, if any, of their own teeth over the years and eventually had to send to the big

Today's cross-country ski enthusiasts would have reveled in the untrammeled beauty and majesty of the Wyoming mountains and the multitude of trails the residents of DuNoir could choose from for a Sunday afternoon of skiing. No noisy snowmobiles polluted the quiet, no long lines of downhillers waited to ride a ski lift — only the soft plop of melting snow falling from a branch, the sharp crunch of skis cutting through virgin snow, and birds singing in the pines.

city for a set of artificial choppers, having first made crude wax impressions of their own jaws. The mail-order plates were seldom comfortable, and they seemed particularly to get in the way during a drinking spree, so they would be taken out and often mislaid. The next morning after a bash would find the woolly-headed men passing around sets of teeth to "try out" to see if they fit well enough to be claimed.

Besides drinking bouts and all-night card parties, the Hacks found fishing a favorite form of relaxation, and a mess

A bright winter's day was as good a time as any for the Hacks to go fishing — through a hole in the ice of one of the upper lakes.

The result of fishing with a hand-held line through the ice was this string of fish

of trout was a welcome change from their beef and pork diet. The small streams, far from civilization, were alive with trout, as were the lakes, where cutthroat and rainbows often reached three or more pounds. Winter fishing meant cutting a hole in the ice and dropping in a hand-held line to tempt a hungry fish.

Game was abundant in the woods. Elk and deer and an occasional bear might be hunted for food, but the moose were special inhabitants of the Wyoming woods, and the men mostly respected their right to live there unmolested.

Not that the moose might not occasionally threaten to molest a human!

Males in rutting season were unpredictable and always given a wide berth, and a cow protecting her young was just plain mean. More than one team of horses and its driver were confronted by one of the majestic animals standing in

A bull elk, weak from hunger, gives up floundering through the deep snow in an attempt to escape the humans who have brought his salvation — hay.

Old Man Winter began his relentless siege of storms unusually early in 1938, catching even the larger game unprepared. The elk herds had not moved to lower elevations before deep snows marooned hundreds of them. The Tie Hacks joined game wardens in sledding hay to the starving creatures near Lake of the Woods.

Going along for the ride when the men took hay to feed the starving elk were these three youngsters, who probably were never again lucky enough to be in such close contact with the wild and beautiful beasts.

Moose were the undisputed monarchs of the woods

the center of the road, head lowered belligerently. Though most encounters ended with the moose walking peaceably away, there were times when the driver prudently backed up his team and went another route. A man afoot was even more easily intimidated.

One winter day Johnson didn't show up for dinner and was found later standing atop a pile of ties, just out of reach of a testy moose pacing around the stack. The searchers shooed the animal away, and a relieved Johnson came down from his perch to eat a cold supper.

An unfortunate forest ranger was forced to spend an uncomfortable night in a tree to escape the wrath of a bull in mating season. By morning the moose had tired of butting his rack against the tree, and the shaken ranger was able to go on his way.

Moose roamed through the camps at will. A bull with good-sized rack munches on a bale of hay meant for the workhorses at this outlying camp.

There were always several moose wintering in the willows along Warm Springs Creek below Headquarters camp, and they would sometimes wander through camp as casually as any of the workhorses they browsed beside. Ungainly animals, their overlong legs and thick bodies make them appear awkward, but where other animals may flounder in deep snow, moose become creatures of grace and beauty as they move smoothly and easily about the woods through the drifts.

Out skiing one winter day, my mother, Agnes Trego, and Betty Dolenc looked down into a valley where four cows and their calves browsed the willows, as well as two bulls equipped with massive racks. The women stood watching and for fun whistled and "moo-ed" at them as they might at cattle. Their dogs, Tippy and Max, took the mooing as a signal to chase the moose. As the dogs ran toward the animals, the cows herded their calves up the other side of the hill, but the bulls turned menacingly after the dogs. Tippy and Max, thinking better of their unwise attack, turned tail and raced toward the women.

Faced with the onrushing bulls, Agnes and Betty quickly took to the nearest tree. Fortunately, the bulls gave up the chase almost as quickly as the dogs had, and the women climbed back down the tree, strapped on their skis, and headed for home.

They never mooed at another moose.

Chapter Three

THE COMPANY STORE

The large, rambling, old log building housing the company store was the equivalent of today's shopping center for the men. It was even open evenings and Sundays for their convenience. They could purchase their rough work clothes, mackinaws, and caulked boots, as well as food and snoose, but no liquor was sold in camp.

However, when a powerful thirst overcame caution, there was often a suspicious run on the supply of vanilla.

Supplies were hauled in by truck two or three times a

The company store was the hub around which activities at Headquarters began. The Tie Hacks' shopping center, the building also housed the company office and U.S. post office. Built before the days of modern refrigeration systems, Mother Nature kept foods and meat cold in the winter; an oversized root cellar built onto the rear of the store was the summer cooler.

Front view of Wyoming Tie & Timber Co. store on a typical sunny winter day. Note woodpile which kept the structure warm. The building to the left is a warehouse and meathouse.

month in the summer — and whenever possible in the winter. The big cellar, cut into the side of the hill at the rear of the store, kept foods cool all summer and nearly frozen all winter. A meat house attached to the side of a nearby warehouse stored big sides of beef and pork. The beef was purchased from ranchers in the valley, and the company raised its own pigs. The meat stayed frozen all winter without the aid of modern refrigeration. Tony Dolenc, storekeeper, would throw a frozen piece of meat on his block — a stump from a big tree — and chunk it with a double-bitted ax. There weren't any neatly sliced steaks or chops. Not knowing how to cut meat properly caused Tony some misgivings when he first took the job of storekeeper in 1936. But then, neither did anyone else.

There was seldom any fresh produce available. Bananas came still clinging to the stock, and occasionally a tarantula

Tony Dolenc, storekeeper, at his cash register. Items most asked for by the Hacks were within easy reach on the well-stocked shelves — "Peerless," "Granger," and "Sir Walter Raleigh" smoking tobacco.

would crawl out from its hiding place near the center of the fruit.

Orders for the cookhouses and families or bachelors living in the outlying camps were put up monthly, filling big wooden boxes. Often the row of orders filled the long store. They were loaded on the truck, or sled in the winter, in the order in which they'd be delivered.

In a 1977 interview, Tony recalled putting up those first orders. "I'll never forget old Carl Pierson. He'd sent his list down to Headquarters, and one of the items on it was 'one can yam' so I sent him a can of yams. Next month, the same thing, 'one can yam' and I sent him another can of yams. Then old Carl came in person and did he give me hell — he wanted *jam* not yams!"

A Norwegian, with not much knowledge of the English language, Carl had spelled it just like he pronounced it.

Tony (in cap, behind sled) oversees the loading of a supply sled. Supplies for the outlying camps were sent out once a month during the winter on large, flat sleds pulled by four-horse teams.

A. B. ("Treg") Trego, secretary and bookkeeper for the Wyoming Tie & Timber Co. He also ran the post office.

By 1942 Trego had become skilled as an amateur photographer. His pictures capture life on the mountain and make up the bulk of the photos in this book.

At one side of the store was the company office and U.S. post office run by my father, A. B. Trego. The arrival of the mail three times a week — Tuesday, Thursday and Saturday — was an eagerly awaited event for me. Besides the fact there was likely to be a box of books from the county library addressed to me, I was often allowed to help sort the letters into the little wooden pigeonholes that made up the post office portion of Trego's office. The mail came by sled in the winter, sometimes by both truck and team in the spring. When the roads were too muddy for the truck to navigate, the contract mail carrier would hitch a team to his vehicle to pull it up the hill.

Always accommodating, the mail carrier didn't mind giving a lift to anyone needing a ride off the mountain. Or back

Neither snow nor sleet — but mud was a different matter. When the roads were too muddy to by navigated by truck, mail carrier Manley Green hitched a team to his vehicle for the thrice-weekly deliveries. In winter mail arrived by sled and team.

on. Many times a Hack who had spent all his money in a Dubois tavern would hitch a ride with the mailman to get back to camp.

The Hacks were free to work their own hours, to come and go as they pleased as long as they cut all the ties in their assigned stand of timber. Each tie was notched with the Hack's own mark, and a record of how many he had cut was kept by the tie inspector, Alfred Olson, then sent to Headquarters where Trego kept the books.

I often sat beside him on a tall stool at the high, slanted counter where he entered what the men had coming in wages. They weren't paid in cash, and there was no payday. When they needed money they would come to Trego and make a draw against their account. The company was their bank. Cash wasn't even needed to buy groceries, since Wyoming T&T issued coupon books for use in the store to

Trego at work in his office. Note the ledger and check register on the high slanted counter.

THE COUPONS IN THIS BOOK ARE
GOOD FOR MERCHANDISE ONLY
AT THE STORE OF
Wyoming Tie & Timber Co.

$5.00

ISSUED TO________________

ISSUED BY________________

No 1460A ON ____________ 19

NOT TRANSFERABLE · NOT GOOD IF DETACHED

"I owe my soul to the company store" was all too often true. Purchases were not made in cash but in coupons issued by the Wyoming Tie & Timber Co. through the office.

Though the company purchased its beef from nearby ranchers, it raised its own pork. Bert Engstrom *(center)* and Albert Edstrom grind oats to fatten the pigs.

purchase food and clothing. The coupon books were also charged against their accounts, so the men had little use for cash other than for their gambling needs or drinking binges.

The mysteries of filing income tax forms or filling out other inscrutable government forms was beyond many of the Hacks' command of the English language, and Trego often assisted them with such correspondence.

One day Frank O'Bosko, a Russian, came into the office with something on his mind. Stamping the snow off his boots and spitting out his wad of snoose, Frank began his story. It seems he was married as a young man in Russia, and he knew his wife had borne him a son after he had left her in the Old Country many years ago to come to America and make his fortune. The years had passed and still no fortune. He lost touch with his family. Now he was old and wanted to see his son. Could Trego help him find the boy?

View of east end of Headquarters taken from hill behind store. at far left is the author's home; next to it is the cabin lived in by the storekeeper and family. The company store is in center of picture. Note woodpile either side of structure — one for the store, one for the office. Warehouse is beside store, and at far right is cookhouse, used in later years mostly for roaring, all-night Christmas parties.

"But not my *woman,* Trego! Only the boy!" he instructed.

Unfortunately for Frank, his sketchy information didn't give Trego enough to go on, and the Russian embassy was less than helpful. Frank never did find his son.

But then, he didn't find his woman, either.

Frank eventually died of the same disease that took many of the old-timers, "whiskey pneumonia." The men would go on such prolonged drinking bouts they would cease to eat, fall victim to a cold or the flu, and die of pneumonia.

Grateful for Trego's help in such correspondence, the men always offered to pay, but he wouldn't accept money for his assistance. Since they were so fond of liquor, they figured everyone else was, and if he wouldn't take money, they'd buy him a bottle of booze. We had one of the best-stocked liquor cabinets around.

They were always eager to treat Trego to a drink if they ran into him in town.

One night in Dubois big Art Smith, the catskinner, spied Trego at a dance. My father, a trombonist, was just finishing a set with the band. Art was weaving when he made his way to Dad's side. Towering over my five-foot seven-inch father, Art beamed and coaxed happily, "Lemme buy you a drink, Trego."

"No thanks, Art — some other time" was Dad's rejoinder. Not to be denied the pleasure of his friend's company, Art persisted but was again refused.

"Dammit, Trego, I said lemme buy you a drink!" Art roared. Reaching down, he lifted my father by the collar and marched him over to the bar.

Trego took the drink.

Photo courtesy Oscar Aspli

Election Day, November 1938. The store's outdoor sign "Post Office, DuNoir, Wyo.," was brought inside, making a wooden table in front of the dry goods section the official voting spot for the Tie Hacks. Election day became a social event, calling for refreshments served by the election board. Note the picture of President Franklin Delano Roosevelt on the wall.

Chapter Four

HARD WORKERS, HARD DRINKERS

The Tie Hacks played as hard as they worked. A *Denver Post* reporter, Bill Hosokawa, described the men well in a 1947 newspaper article:

"They were lusty men who loved the pungent woods and nurtured a fierce pride in their skill. It was their creed to work as hard as their rock-like muscles could drive them, and when it was time to play they brooked no inhibitions.

"At a tender age, the Tie Hacks learned to cultivate the three Bs — booze, bawds, and brawls. They grew into men blessed with lead-lined stomachs, and cursed with an inextinguishable thirst for alcohol. They mauled strangers and each other for the sheer love of mayhem.

"They brought to the Wyoming forests the wild abandon of their Viking forbears, and their prowess with broadax and bottle fathered legends that well might have been perpetuated in Beowulfian saga."

The Hacks would work for months on end without leaving the woods. Then an occasion would arise for celebration, or their thirst would get the best of them, and they'd be gone — sometimes for weeks or until their money ran out. More times than not the latter would happen before the thirst was quenched, and Trego would get a phone call from one bartender or another in the nearest town of Dubois, wanting to know whether John or Ole or Swenson was good for another $10. Like as not Trego would consult the books and determine, well, all right — but no more! Send him home when that's gone.

Once a drinking bout didn't end soon enough for Ole.

Photo courtesy Kersten Shepardson

A group of old-timers, taken in 1926. *Left to right:* Bill Murray, Chas. A. Johnson, John Gist, G. B. Dole, George [illegible]ll, Gust Pederson, John Toomey.

Ole Erickson

John Berglund

Photo courtesy Lydia Olson, circa 1928

Drinking was a major recreation for the Tie Hacks. Drinking bouts lasted several days to several weeks — until the money ran out.

Back in camp broke, needing to go to work but so shaky he couldn't hold an ax, he knew he needed another drink to sober up. Desperate, Ole raided the cookhouse while the cook was outside, drinking every kind of extract he could find that might have a little alcohol in it — vanilla, lemon, maple flavoring. He even polished off a bottle of green cake coloring.

The latter was his undoing. With the shakes somewhat subsiding, Ole went to work. But the "cure" brought on a monumental case of diarrhea. That night in camp the Hacks all roared about the "whole hillside painted green" where Ole had been hacking ties.

There is a narrow switchback on the road leading to DuNoir known as Tie Hack Dump. It was named for an

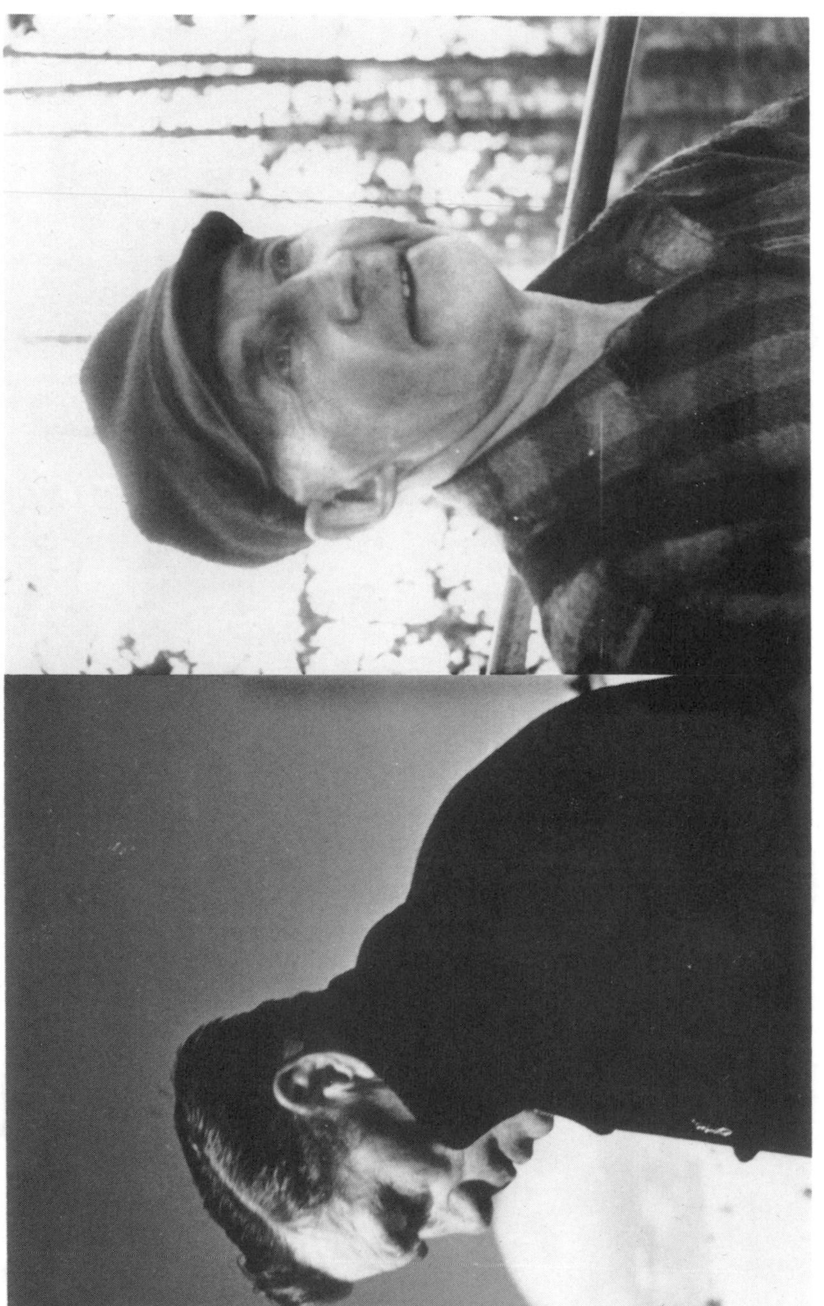

Ted Berg

Sorn Pederson, blacksmith

Art Smith, catskinner

Arapahoe tie piler in the Riverton Yards

Photos taken in mid-'20s, courtesy Lydia Olson

Sorn Pederson, blacksmith, at work, watched over by Martin Olson, woods boss

Photo courtesy Lydia Olson

Seemingly impervious to the cold, Bill Murray, tie hauler, sits atop his snowy load of hand-hewn ties, ready to move them to the edge of a stream bank where they will be stacked for the winter. The men wore "packshoes" such as Murray's to keep their feet insulated from the cold and snow.

anonymous drunk Hack whose car ran off the road, rolling end over end down the steep slope until it crashed into a tree. Rescuers found the driver sitting amidst the wreckage, smiling blearily, his thumb securely stoppering his corkless whiskey bottle. His car was wrecked, but he was all right. Most important, he hadn't lost one drop of whiskey!

The need for liquor was sometimes so overpowering that a man would go to any lengths to get it. The story is told of Little John Pierson who lived in a bachelor camp on South Fork. Little John once skied to Dubois twice in one day to pack liquor back to camp, a journey of over twenty miles each way in deep snow. And the return trip was all uphill.

Loneliness and maybe frustration over the forty-mile round trip evidently drove Little John to despair. One bleak winter day in 1937 he took a high-powered rifle, placed the barrel in his mouth, and pulled the trigger. A shocked community notified the county coroner, who arrived the next day in Headquarters from his office in Lander, one hundred miles away. After traveling another eight miles into the mountain fastness by bobsled, he began his investigation. The men were as puzzled as the coroner about Little John's reason for ending it all. A newspaper account of the event stated the deceased was "referred to by one and all as a good workman and a peaceable man." The coroner persisted with his queries until the exasperated woods boss, figuring enough was enough, delivered the final appraisal: "He knew of no reason why Pierson would want to commit suicide, unless he was just tired of living."

It wasn't the coroner's year. He had to make that same miserable trip back to DuNoir and the upper camps only a month later when an avalanche trapped and killed another Hack.

As a matter of fact, it wasn't a good year for the Hacks either. Another man was killed. In January Emil Peterson

Sawmill crew shuts down for lunch. Leo Moon, Bill Armitage

died of injuries received when a boulder rolled onto him. He was buried in the Wyoming Tie and Timber's burial plot in the Riverton cemetery. Another taciturn Swede, his newspaper obituary revealed the loneliness of his existence: "Little is known of his life's history. The fact that he had made his company insurance policy payable to the Employees' Hospital Fund leads one to believe he had no living relatives."

In all the years of its existence the company held an enviable safety record of only two men killed on the job, one of them being the lonely Emil.

Though most of the Hacks were Scandinavian, there were two brothers, Bill and Louis Brown, who were black.

Photo courtesy Oscar Aspli

On Camp Creek Oscar Aspli *(left)* and John Velseth teamed up as partners in hewing ties. They stand on top of a day's cut.

Tools struck into a completed tie, four Hacks take a lunch break. *Left to right:* Big John Olson, Oscar Aspli, Ole Tverdahl, and Lars Kvam.

Photo courtesy Oscar Aspli

In 1928 the newly completed Headquarters camp on Warm Springs is ready to be moved into. Union Pass Trail winds up the mountain across the creek *(left)*. The logs for the buildings were cut from the top of Union Pass.

Bill had a ear for language and learned to talk Swedish as well as any of the Swedes themselves.

One time in the early 1920s when a trainload of new recruits from the Old Country was arriving, Bill was sent downcountry to pick them up. The neophyte Tie Hacks were taken aback when this strange black man spoke to them in their own language.

"*You're* not Swede!" they protested. But Bill shot back: "If you stay here in this country as long as I have, you'll turn black, too!"

They believed him. A few days later one of the new men took the woods boss aside and, motioning toward Bill, asked in a quavering voice how long it was going to take before he turned that color, too.

Photo courtesy Oscar Aspli

Spending a lonely Christmas thousands of miles from his family, a young Oscar Aspli climbed to the top of Lookout Mountain Christmas Day, 1928. A friend snapped this picture, which Oscar sent to his family in Norway.

Bill had not only a proclivity for languages but also for fighting. A powerfully built man, he was always eager to engage in one of the joyous, bare-knuckle battles that occurred sometimes among the men. Whoever won was respected as top dog of the crew — until a new challenger came along and disputed the title.

One of the longest and most historic battles recalled by old-timers was between Bill and a young Swede named Pete.

These two hammered at each other in a marathon event that lasted more than a week. They fought before breakfast and after dinner, each bout ending only when one or the other was knocked out. The end came at last when Brown didn't get back up. In fact, he went out of his head not too long after this defeat and had to be carted off to an asylum for a while. Once his brains had unscrambled themselves he returned to hack ties for several more years before the cumulative effect of many such bouts signalled his demise.

Oscar Aspli was a young Norwegian of twenty-six in 1926 when he kissed his wife Dorothea and baby daughter Ingebjorg good-bye and sailed to America to find work. Working his way west, in the Wind River country he met a man who was leaving the woods. Oscar bought his bedroll and tools — a chopping ax, a broadax, and a peeler. He learned to hack ties by the simple expedient of picking up that broadax and "yust doing it. If I didn't know how to cuss before, by golly, I sure learned it then," he says today. Even so, he hacked seventeen ties his first day at work.

Ten years later he had saved enough money to bring his family to America. When he met them as they came off the boat his bewildered wife couldn't understand a word he said. He now spoke a peculiar jargon of Norwegian, Swedish, and English.

I remember the first day his daughter, Ingebjorg, came to school at DuNoir. She was a beautiful girl with long black hair in braids down her back who charmed us all with her smile, although she couldn't speak a word of English. It

Alfred Olson, tie inspector and drive boss. Younger brother of woods boss, Martin Olson.

Photo courtesy Oscar Aspli

Ingebjorg Aspli ready to ride her horse, "lutt," the five miles to school.

Photo courtesy Oscar Aspli

Bringing home meat for the winter, Oscar Aspli uses a skidding horse to help with the heavy work.

didn't take long before she sounded just like all the rest of us, without a trace of her Norwegian accent.

For five days a week Ingebjorg was a boarder with a family in Headquarters during the school months. But the year she was in the eighth grade the Asplis were living in a camp only five miles away and she was able to ride her faithful and dependable horse, Mutt, to school every day until late November when the snows became too deep.

Oscar had been one of the building crew who put up the log buildings for the new and final Headquarters camp on Warm Springs Mountain back in 1927. (Old Headquarters had been built on DuNoir Creek, and the U.S. government had named the post office DuNoir. When the timber in the DuNoir Creek drainage was exhausted, the move was made to Warm Springs Creek drainage, but the post office name stayed the same, and new Headquarters camp also became DuNoir.)

The logs for the buildings came from trees on Union Pass Mountain, just across the stream from Headquarters. Oscar would ski home in the evening with his cross-cut saw on his back. But in the morning it was a different story. To assist in that tedious climp uphill, he and the others would hang onto the horses' tails and let the animals pull them up.

Long retired from the woods, Oscar lives in Riverton today. Looking back on his days as a Tie Hack, in his still-broken English he recalls with relish the story of a fishing jaunt three of them took, walking several miles into the wild and beautiful area called Seven Lakes:

"Squeak Axel and Andrew Bloom and me, vel, ve vas goin' fishin. Andrew had been on a toot, y'know, but vas bound he vas goin' vid us. It took us all day to get to Seven Lakes cuz climbin' up dat hill ve called the Turntable, ol' Andrew, he yust played out. He vould lay down and I tink, 'By golly, he gotta get up sometime,' so I'd go back to him, y'know, and coax him a liddle bit and he'd get up and go a vays and den he'd lay down again.

" 'Leave me alone!' sez he, 'I vanna yust die.' Vel, y'know, ve couldn't do dat, so I finally got him up again, and ve finally made it to the top. Ve had a liddle shack der at the lake, y'know, made out of spruce boughs vid a tar paper roof that had a stovepipe goin' up through it.

"And y'know, ol' Andrew, he yust *laid* in der two days!"

Chapter Five

OLD TIE PIN

The man responsible for the success of the tie-making operation was Martin Olson, woods boss for over thirty years. To the men who worked for him he was "the Tie Pin."

Martin came to America in 1897 at the age of seventeen from Namsos, Norway, already a seasoned lumberman who had been on river drives and had been a deckhand on lumber schooners plying between his native town and England.

When he arrived in frontier Wyoming he found his first job at a tie camp near his uncle's homestead at Elk Mountain in the southern part of the state. Here he borrowed a nine-pound broadax (average weight was seven pounds) and became a Tie Hack, hewing twenty ties that first day.

He learned all aspects of the trade — hauling, cutting, driving — and before long was recognized for his skill in working with men.

In 1916 he was called to the upper Wind River country to help salvage a fouled-up river drive. He accomplished this task and was asked to take over the job of woods boss, replacing a man by the name of Flint, who was a stern teetotaler. Flint had imposed a strict rule of no liquor for the crew. By the time Martin took charge, the men were in a mutinous mood.

A man who appreciated a drink or two himself, Martin saw no problem in changing the rules. The men soon learned he was fair but no milksop. He had no hestitation about tangling with a logger twice his size if he felt the man

Martin Olson, the Tie Pin himself, woods boss for over fifty years

Martin in the office, looking over the accounts

needed a whipping. Years later a co-worker said of Martin, "He had a way of getting the best out of an ornery crew — they would have carried the season's cut out on their backs, with Martin in the lead, if it had been necessary."

He could be friend, father-advisor, even confidant, but always boss. "Be like them, but don't sleep with them" was his philosophy.

His other genius was in knowing and loving the woods. Trego wrote years later about this stocky, taciturn man:

"Martin was a quiet man, keen and able. He never wasted words. Life in the woods was rugged, and it took a special kind of man to run a successful timber operation. I was always amazed at Martin's knowledge of timber and how he could walk through a drainage area and predict almost exactly how many ties they would be able to market. Over the years his calculations proved out so exact they were hard to believe.

"Martin also had a special way of handling men. He was tough when the situations called for it but never abused any man who worked for him. He left them with complete freedom to work or go fishing as they wanted. If a Tie Hack wanted to get drunk and blow his wages, that was his own business so long as it didn't interfere with anyone else. If a whole crew went on a party they were given so long to get it out of their systems and then were told to either get back to work or 'roll their beds.' "

Three years after Martin's arrival in the Wind River country, another man came into the picture: Ricker VanMetre, who was to become the president of the company. VanMetre later wrote of that first meeting:

"In 1919 a group of Denver bankers with money invested in the Wyoming Tie and Timber Co. went to Chicago seeking a man to manage the company and if possible salvage something of its dwindling assets. I was sent to Wyoming to survey the situation.

"In October, 1919, Martin guided me over much of the potential cutting area of upper Wind River as well as DuNoir and Warm Springs creeks. The timber was there in abundance — but to me it looked impossible to get out. The river drive was the only way to transport the ties to Riverton, and the cutting area accessible to the Wind River was limited. It seemed scarcely feasible to get ties out by the tributary streams.

"But Martin said it could be done, and I had a hunch that here was a man whose judgment could be trusted.

"It was the best hunch I ever had.

"On the strength of his word, the word of a man I had only just met, I accepted the management of the Wyoming Tie and Timber Co."

It was indeed a good hunch. The working relationship between the two men was to last until 1947 when the company was sold, and they remained fast friends until VanMetre's death.

With VanMetre in charge of the business end and Olson in charge of the woods, the company flourished. Martin was

Over 300,000 ties clog Warm Springs Creek, held in check by the boom stretched across the mouth of imposing canyon in the center background.

right; he could get the ties out of the remote areas. Where the tributary streams were too small or too littered with rocks to drive ties, he built flumes — great V-shaped wooden troughs — down which the ties were floated.

He built flumes in places engineers said it couldn't be done, including the longest, Warm Springs, which was built in 1928 for $63,849. Warm Springs started three miles east of Headquarters on Warm Springs Creek and carried the ties nine miles through a rugged, perpendicular-sided canyon. The flume emerged on the Wind River some six miles above the town of Dubois.

A remarkable feat of engineering, this flume and its cat-

The boom open, the ties are funneled through to the mouth of the Warm Springs flume where they begin their nine-mile float down the canyon, to pour out with a mighty splash into the Wind River. From that point it's still a 100-mile trip to the tie yards in Riverton.

Photo courtesy Lydia Olson, circa 1928

Martin's smile shows satisfaction with the completion of a job well done — the construction of the nine-mile Warm Springs flume in 1928.

Photo courtesy Lydia Olson, circa 1928

Another view of the Warm Springs flume when first constructed in 1928. Note cables anchoring flume to canyon wall.

walk clung to sheer walls of the canyon, and in one place followed the stream underground into a large limestone cavern known locally as the Natural Bridge.

Though constructed by an engineering firm from Spokane, Washington, it was built as Martin conceived it. The men would build a section of the flume at a time, turn water into it, and float materials down for the next section. They worked a year to complete the structure.

It was last used to transport ties in 1942. Today most of the flume has been destroyed by rockslides and the elements, though the portion that travels through the Natural Bridge has been captured for all time, preserved by the precipitate that drips constantly from the roof of the cave. Stalactites, stalagmites, and the wood-turned-stone flume create an eerie world of fantasy.

Canyon Creek flume was probably the steepest timber flume in the world, at one point plunging down an unpre-

Warm Springs flume was built to transport the ties through a canyon too steep and stream too rocky for driving. It took a year to build.

Photo courtesy Lydia Olson, circa 1928

At one point the canyon walls were so steep the flume and its catwalk were suspended by steel cables.

The flume traveled through a great underground cavern filled with stalagmites, called the Natural Bridge.

Photo courtesy Lydia Olson, circa 1930

Inside the Natural Bridge it is eternal winter, a fantasyland of ice and crystalized formations.

Photo courtesy Lydia Olson, circa 1930

Looking east from the exit of the Natural Bridge at the lower end. Flume can be seen along canyon wall in background.

cedented 45 percent grade. Only 2,200 feet long, it dropped 1,800 feet in its descent to the Warm Springs flume. One year an attempt was made to dry-flume the ties down Canyon Creek, but the plan was quickly abandoned when it was discovered the ties shot down so fast the friction set the flume on fire.

Martin, as has been noted, was as fond of drink as any of his Tie Hacks, and it was he who always concocted the annual Christmas punch by squeezing a few lemons into ten-gallon cans filled with whiskey!

He was a periodic drinker. His wife Lydia recalls with a rueful smile, "I never could tell when he was going on a toot." When he did drink, it was never because of a problem, however.

"*That's* no way," he said. "You can't work out your problems with booze, they'll just be worse." No, when Martin went on a drunk, he went to have fun.

It was Martin's task to determine the precise time to start

A dramatic sight, the flume leaves the mouth of the cavern for the final leg of the journey through Warm Springs Canyon.

Photo courtesy Oscar Aspli

The square trough in center of picture leading into the Canyon Creek flume was a feeder trough, bringing water from the creek to replenish the flow in the flume.

Photo courtesy Lydia Olson, circa 1930

Youths seated on the edge of the Warm Springs flume look across the canyon to the bottom of Canyon Creek flume and its water trough. Note water pouring from sides of water trough.

During the drive, ties coming down the Wild Cat flume plunge into the dam pond to join other ties in the float down Warm Springs flume.

Possibly the steepest lumber flume in the world, Canyon Creek flume was constructed on an awesome 45 percent grade at one point. This flume fed directly into the Warm Springs flume.

The Wind River Valley seen from the top of Warm Springs Canyon, Ramshorn Peak in the distance. The flume has traveled over eight miles through the canyon at this point, and the precipitous walls give way to more accessible terrain.

Photo courtesy Shirley Daniels

View of the dam showing gate lowered across mouth of the Warm Springs flume. Wild Cat flume winds down the mountain (*right*).

Looking west toward the Warm Springs dam after the annual drive is completed. Not needed for another year, the log boom across the mouth of the dam has been removed. Coming down the left side of the photo to empty into the dam pond is another of the Wyoming Tie & Timber Co.'s longer flumes, Wild Cat. The main Warm Springs flume leaves the dam at the right of the picture.

Photo Courtesy Arden Coad

Today only a skeleton of the Warm Springs flume can be seen, and the remaining sections are fast deteriorating.

the drive each year. Lydia explains: "He knew by the looks of the streams and the amount of snow left in the woods. When the water is rising a stream is high in the middle. When it drops it goes hollow-looking, then your ties don't get hung up on the banks.

"There is a place up behind our old house, a ridge, where he could look over the valley. We'd go there together in the spring, and Martin would tell me, 'Now, it's driving time.' He could tell by the amount of snow that laid on this ridge. In all those years he ran the show, we never had a year he couldn't drive."

Much of the backwoods of this high country is open for forest service and tourist travel today because of the roads

These treacherous switchbacks opened Little Warm Springs Canyon for a new tie-cutting operation. The engineer hired to survey the road told the woods boss it couldn't be done, so Olson did it anyway, by walking in front of the bulldozer and explaining to the driver where he wanted the road cut. President of the company, Ricker VanMetre, looks over the completed project.

Another view of the Little Warm Springs Canyon road, Wind River in the distance.

built by Wyoming Tie and Timber Co. Martin more often than not was his own surveyor. In one instance he decided the terrain was too steep for his unschooled abilities; he hired an expert to lay out the Warm Springs Canyon Road. When the surveyor arrived on the scene and looked at the task before him he shook his head and vowed it couldn't be done.

Disgusted, Martin discharged the man and went ahead and did it his way — by walking in front of the bulldozer and explaining to the operator where the road should go.

Martin retired in 1947 after fifty years in the woods. A sign erected over the gate at his and Lydia's new home in Dubois proclaimed to the world "Dunloggin." He put in six productive years after that as a county commissioner in

After fifty years in the woods, Martin retired, and he and Lydia built this new home in Dubois, designed and constructed as much as possible like a "quality" home in Norway. Olson extends his hand in greeting to old friend and co-worker, Ricker VanMetre. Lydia stands beside her husband, and VanMetre's wife, Louise, stands underneath the sign proclaiming to the world Martin is "Dunloggin."

Photo courtesy Lydia Olson

On his eightieth birthday Martin hefted a broadax to demonstrate how he had hacked ties in his salad days.

charge of something he knew best and Fremont County needed most: road building.

His drinking sprees tapered off but didn't end. In 1952, when Martin was seventy-two, my father wrote there was a flu epidemic in the county. "Even Mrs. Olson is in the hospital with it. Martin is too well sterilized to get the flu. He has been staying sober the last couple of years, but about a week ago he came to town [Riverton] and really had himself a time.

"Yesterday I saw him go by on Main Street, two of the Tie Hacks with him — Charley Nordlund on one side and old Brady on the other, all tighter than ticks. If they had been wearing crowns they wouldn't have been more like three happy kings walking down the street. Today I saw Martin and he was sober as a judge, listening to a symphony concert at the auditorium. . . . Quite a fellow!"

On his eightieth birthday he was toasted at a giant Wyoming celebration at Battle Mountain, where he posed with broadax in hand astride a felled tree, to prove he still knew how to hack ties.

Stricken with cancer and nearly blind, he died at age eighty-nine, just a few months after Trego died of a massive stroke in 1969.

Ricker VanMetre died three years earlier, in 1966 at age eighty-four, and the other man of the working triumvirate, Billy Mac, died at age seventy-seven in 1953.

William H. ("Billy Mac") McLaughlin, assistant superintendent for the Wyoming T&T, ran the tie yards in Riverton, bringing the annual harvest to a successful conclusion. He was with the company thirty-five years, having signed on even before Martin, in 1913. It was Billy Mac who designed a system of islands, fingers, and booms on the Wind River in Riverton, diverting the main channel to avoid loss of ties during high water. He also designed and built the yarding conveyors which brought the ties out of the water and into the yards.

The working triumvirate who shepherded the Wyoming T&T through its perilous beginnings to its successful conclusion: Martin Olson, woods boss; William H. ("Billy Mac") McLaughlin, superintendent of the Riverton tie yard; and Ricker VanMetre, president of the company.

Chapter Six

THOSE MAGNIFICENT DRIVES

The end of the river drive every summer was justifiable cause for celebration. The men had spent weeks on the river, guiding errant ties back to the main channel, pulling stranded ones off sandbanks, pushing them over Diversion Dam while standing precariously atop the spillway. The men's lives during those weeks were not unlike that of cowboys herding truculent cattle on a big roundup. Except for one big difference: Tie Hacks' feet were never dry. The first day out they punched holes in their boots to let the water run out freely. Constant wet feet was the cause of a common painful ailment of the river pigs — "squeak heel."

A strange dichotomy exists in squeak heel. Though the wetness causes it, the Achilles tendon itself inside the skin dries out and "squeaks." The Tie Hacks sought to ease the discomfort by applying great gobs of vaseline to their wool socks in an attempt to keep the heel lubricated.

Just as cowboys had to contend with sudden stampedes, the Hacks fought sudden cloudbursts which could send the waters of the Wind River over its banks, scattering ties from here to Sunday. Each stray had to be located and lugged back to the river on a strong shoulder.

A single tie could hang up on an old, half-submerged snag in the river. Then all the ties coming behind it would pile up, one on top of another, creating a massive jam. Looking down on such a tie jam from a highway viewing point, it seemed for all the world as though a giant playing pick-up sticks had tired of the game, leaving his oversized jackstraws all in a heap.

Freshet over, the year's crop of ties begins to fill Warm Springs Creek on the way to the dam. Moose often grazed the willows, unconcerned with the unusual activity in their territory.

The first drive camp of the summer was always set up only a short distance from Headquarters camp (cabins in background), while the men worked the ties behind the boom at Warm Springs dam, poling them one by one into the flume.

After a long day in the river the men would climb aboard a waiting truck for the trip back to that night's camp, where they slept in bedrolls on the hard ground, sheltered by teepee tents. The drive cook and his flunkies moved their converted truck-chuck wagon with the drive as it advanced down the river. By the time the day ended for the soggy crew, huge quantities of food were ready for them. There were always at least two kinds of meat, potatoes and gravy, vegetables, and thick slabs of crusty homemade bread — all devoured silently by the bone-weary workers. The daily feasts attracted curious tourists and knowing locals alike. Anyone was welcome to pick up one of the enameled metal plates and queue up to be served heaping portions from the big, black dutch ovens.

By the time the tired crew had finished contending with blazing sun, sudden downpours, swarms of flies and mos-

The mess wagon was protected by a big canvas which stretched out to form a tentlike, open-air kitchen, from which the drive cook and his flunkies fed the hungry crew.

Photo courtesy Shirley Daniels

Food was cooked in big iron dutch ovens placed over hot coals in an open trench. More hot coals were heaped on the lids of the ovens, and the food came out steaming hot and mouth-watering. Bread was baked daily on the drive in the same fashion.

Oncoming tie splashes into a snagged tie in the flume, and the crew gets a shower trying to straighten out the tangle. Here, against the Wyoming badlands in the background, the nine-mile Warm Springs flume crosses a short stretch of level terrain before emptying into the Wind River.

quitoes, and perpetual wet feet, they were *ready* for a celebration. When the last tie was safely herded into the yards in Riverton the men headed for the nearest bar, and the townspeople resigned themselves to the inevitable brawls to follow.

Heavy with understatement, an item in the September 23, 1937, issue of the *Riverton Review* noted: "The drive crew let it be known the drive was completed, and the event was celebrated in the usual manner with hilarious entertainment and frequent refreshments."

That year there was more cause than usual for celebration. The vagaries of Wyoming weather had played havoc with the progress of the drive. It had started May 1 in the woods with the creek drives down to the Warm Springs

Photo courtesy Lydia Olson

Ties held in check behind a log boom clogged the horseshoe bends of the upper Wind River near the site of the first Headquarters camp in the mid '20s. The Wyoming T&T began its operations on Sheridan Creek, nineteen miles northwest of Dubois, in 1914.

Photo courtesy Oscar Aspli

Close-up of the log boom holding ties in check on the 1925 drive

Photo courtesy Lydia Olson

Bedrolls and tents loaded on the wagon, breakfast over, morning mists still rising off the mountains, and it's time for the men to move on down the river. Note the black dutch ovens strung out in a row through the sagebrush. Though this photo was taken in 1918, hungry river crews were still being served bounteous meals in similar fashion on the final drive in 1946. Today, Dubois firemen use the same dutch ovens to feed the public at an annual Buffalo Barbecue.

By the 1930s drive crews were riding to the river in trucks, carrying their long pike poles used to prod the ties away from banks and into the main channels.

Crew enters the river to begin day's work

flume. Hampered by excessive ice in the creeks and unusual fluctuations in the water flow, it had taken five months for 108 men to complete the relatively small harvest of 377,000 ties. Peak year for the Wyoming T&T was 1927, when 700,000 ties were yarded in Riverton for the Chicago and Northwestern Railroad.

There was only one year when the Tie Hacks didn't make their annual "walk to Riverton." In 1933 union men

Photo courtesy Lydia Olson

Powder monkey Brady Kjelmo *(left)* and fellow worker Casper Benson pause for the camera while breaking out a landing on a tie drive in the early 1920s, pike poles in hand. Toothless Brady had a low raspy voice that sounded as though he had a perpetual case of laryngitis. He had an astonishing thirst and an equally astonishing way of walking, sober or drunk, that defied gravity. He tottered, first on one foot, then the other, arms flapping like ineffectual wings as he sought his balance. He was the powder monkey — the man who picked the exact position for the dynamite charge which would loosen a hopeless tie jam in the river. Onlookers lining the riverbanks would watch breathlessly while Brady set the charge, then made his perilous journey back across the ties, tottering slowly from side to side. His timing was superb — he made it to safety always a second before the dynamite went off.

Innocent-looking puffy clouds drifting across the Wyoming sky give no evidence of the cloudburst a few hours earlier which caused the waters of the Wind River to spill over its banks, then as quickly fall again into the normal channel. Ties stranded on sandbars or the banks had to be pulled back into the water by the men working with their pike poles.

When ties were stranded too far from the river to be pulled back into the water they had to be hefted to the shoulder and carried back.

A white teepee city, nestled along the riverbank at the edge of the well-known Red Cliffs along Highway 287 east of Dubois, was a signal to valley residents that the annual tie drive was under way.

from the city were at work in the woods, urging the Hacks to organize. While opinions were divided, feelings ran high. Eventually the organizers tried to force a union, with the backing of those Hacks who were in agreement. Martin and the company held out, however. "I'll close it down first," Martin threatened. And he did. There was no tie drive, and efforts at unionizing were defeated. The winter's cut was held over until the following year and sent down the river with the 1934 harvest.

World War II created an increased demand for railroad ties. At the same time the armed services depleted the ranks of able-bodied men left for tie-driving. In 1942 the drive crew consisted of only about one-quarter experienced Hacks,

In the final years of the river drives the ties were trucked to the Wind River rather than flumed. Ties were stacked along the banks all year, to be pushed into the water in midsummer. The decks were three tiers deep and about twenty ties high and stretched for a quarter of a mile along the river.

Assisting the crew in breaking out landings is the company bulldozer driven by Art Smith

Photo courtesy Oscar Aspli

Two views of the work at Diversion Dam, tie drive of 1925

A special system of booms, fingers, and islands was designed by Billy Mac (William McLaughlin) to contain the ties as they approached Riverton.

The ties were snaked out of the river and fed onto a conveyor chain which passed through this inspection hut, where each tie was checked to be sure it met specifications.

one-quarter Indians, and one-half eighteen-to-nineteen-year-old boys just out of high school.

Wyoming T&T advertised it would pay board and room and eighty-five cents per hour for men to work the drive. The pay attracted many Riverton High graduates, most of whom were waiting to be called up to the draft. The drive sounded like an easy way to make some money in the meantime. Bob Peck, now editor and publisher of the daily newspaper, *The Riverton Ranger*, was one of the young men who bought a used pair of hobnail boots (too big) and went along with his friends for a summer on the river.

After the motley crew was assembled and drive boss Alfred Olson looked it over, the company reconsidered its generous offer and decided arbitrarily to pay the youngsters only seventy-five cents per hour. Early the next morning

The yard crew snagged each tie as it came along the conveyor, pulling it off onto a stack alongside the chain.

Two other views of yard crew at work in Riverton tie yards

All the ties were tallied by the clerk, Bill McLaughlin, son of Billy Mac

when it was time to roll out of the tents and head for the river, no one stirred, including the Tie Hacks. They had decided to support the boys' expectations of the higher wage.

"It was my introduction to labor negotiations," laughs Peck today. Alfred came around to find out why the men refused to work. When told, he pondered for a minute and said, "Well, we gotta go talk to the Big Boss." (This was Martin, Alfred's brother.) Peck and a college lad were selected to plead their case, and Alfred drove them silently to see Martin, who was sick in bed with the flu.

Hearts pounding, the young men entered the boss's bedroom and stood respectfully at the edge of the bed. Even with a runny nose Martin's stern visage was intimidating. Peck stated why they thought it was unfair for the company to renege on its earlier promise. Unsmiling, Martin heard them out, blew his nose, and announced calmly that of course they were right, and the company would pay tried and untried alike eighty-five cents per hour.

Wearing his ill-fitting boots (which helped bring on a classic case of squeak heel) Peck's first job was "mudding" the Warm Springs flume. Built nearly twenty years earlier, it was the last time the old flume was used. Its once green lumber had long before shrunk, and the trough leaked worse than the proverbial sieve. To cut down this problem, quantities of dirt were shoveled into the flume at the start of the drive. The water moved the resultant mud along, and as it sank into the bottom it sealed up the worst of the leaks.

By the time the fluming was completed and the walk to Riverton on the river was in full progress, the teen-agers were

Tie Hacks devoted fall and six months of winter to tie making. Each Hack cleared his own strip road, felled his trees, and hacked them into seven-inch by eight-foot ties. He stacked them along his strip road for the hauler to pick up.

learning that the experience wasn't going to be a lark. Their days started early and came to an end only after a nine or ten-hour day of standing in water, sometimes up to their necks, pushing stubborn ties along the channel.

Alfred was everywhere — up and down the riverbanks, checking the lead men, the tail donkey, a tie jam. He seemed to appear out of the blue when least expected. Peck wrote fondly of him after his death in March 1979: "To encounter Alfred coming down the riverbank to check a jam-up brought a lump in your throat. Short, wiry, lean, wearing his pants tucked in his boots, a broad felt hat shading bright eyes, and a brush mustache, Alfred's appearance triggered furious activity."

The temptation to slack off must have been a constant one for the teen-agers. Several simply quit along the way. Peck was determined not only to finish but to prove he was as able as any Tie Hack. When it was necessary to break out

The haulers loaded the ties onto their sleds and skidded them to the creek banks, where they were stacked in "landings" to be pushed into the water after the spring thaw.

Photo courtesy Faye Olson

Alfred Olson, tie inspector, atop a landing of ties

Photo courtesy Oscar Aspli, Circa 1925

The lead or "cribbing" crew ready to go to work on the 1925 river drive. These men were sent ahead of the ties to crib off any side channel or ditch where the ties might wander.

a tie jam, the more seasoned of the crew knew the most effective method was to work ties out of the pile-up from the rear, not the front. On his first jam, Peck was standing in water up to his armpits, struggling to loosen a tie in the front of the jam, when an old-timer waded up to him and remonstrated: "You're gonna use yourself all up if you don't learn to f--- the dog." What he meant was for Bob to move to the rear and *look* as though he were busy.

By nighttime the exhausted youth were ready for little more than dry socks and a night's sleep. They watched dumbfounded as the hardened Swedes vied against each other in evening tie-lifting contests, seeing who could lift the largest number at one time.

"I don't know how they did it — those waterlogged ties weighed a good 150 pounds each!" says Bob, shaking his head at the memory.

You could walk from the Warm Springs Creek dam upriver three miles right in the center of the creek just before the drive began. It was solid with ties waiting to be floated down the flume.

When the drives were completed and the men had celebrated, the Tie Hacks returned to the woods and began the tie-making process all over again.

Fall and winter months were devoted to cutting the ties and stacking them in landings on the creek banks. Hand-hewing gradually gave way to sawn ties and winter-sledding to truck-hauling over the years. But for those who cut and hacked ties by hand, the process was always the same. They were each assigned a strip of timber by the tie inspector. The trees had all been marked by the Forest Service for cutting. Those were the days of selective cutting rather than clear-cutting, and only trees stamped with the U.S. Forest Service insignia could be harvested.

Before he could fell his tree, the Hack usually had to shovel down through the snow to the base of the trunk. If the hole was deep enough it meant he would have to cut steps into the snow in order to make a quick exit after he had sawn through the trunk and before the tree fell.

But his job wasn't just felling the trees and hacking the ties. The Hack also was his own road maker. He first had to cut a swath through his strip of timber wide enough for a team and sled. When he had finished hacking the ties, he shouldered each one and stacked them along his strip road so the hauler could come along and pick them up on his horse-drawn sled.

The hauler in turn stacked the ties in landings along the creek banks. Come spring and the thaw, the ties were pushed into the creeks or diverted into the smaller flumes which honeycombed the backwoods to start their journey down to Warm Springs Dam, where they were contained behind a boom stretched across the stream at the mouth of the canyon. From that point they were poled one by one into the big flume for a nine-mile float through the canyon, exiting into the Wind River with a mighty splash.

Once again the river drive was under way.

Chapter Seven

THE COOKHOUSES

A Tie Hack could put away more food in a day than some families eat in several. Martin Olson once figured it took over a ton of food to feed each man for a year.

Meat was the mainstay of the diet and appeared in large quantity at every meal served in the camp cookhouses. Since food loomed large in the men's lives, the role of cook was looked upon as special. If a newly hired cook didn't live up to the men's standards they became surly.

For a couple of years, my aunt, Bertha Millard, was one of the men's favorite cooks.

She agreed to try cooking for the men at a time when she knew hardly anything about what would be expected of her. The only advice given her was *not* to put out napkins for the men.

"Can you imagine! I was used to cooking for fancy society ladies in a tearoom in Grand Island, Nebraska, and here I was in Wyoming, preparing three meals a day for thirty hungry woodsmen." Now ninety-two years old, she laughs as she recalls her first few days.

"That first meal, I didn't have any meat to cook because a bear had raided the meat house and taken it all. My first introduction to camp life was watching the men catch that bear — it was a big one — in a trap in the middle of the little creek there in camp."

Bert soon learned to put out lots of lunch makings at breakfasttime so the men could prepare their own lunches to take into the woods.

But not that first day. As cook she figured it must be her

Circa 1942

Bertha Millard, the author's aunt, came from a ladies' tearoom in Grand Island, Nebraska, to cook in a rough woods camp for thirty men.

duty to make the lunches, and she insisted on preparing one for a man named Schmidt. Schmidt protested but to no avail, and Bert made up two dainty tea sandwiches for him.

He didn't say a word, just took the proffered lunch and apparently ate it — but he never requested a repeat. Bert learned from watching the other men make their lunches that tea sandwiches were not acceptable fare — not even good appetizers.

The tale is told, in fact, of Big John, a Hack of legendary size with appetite to match, who packed his daily lunch in a wooden orange crate.

Schmidt, the hungry victim of Aunt Bert's first day, was a German alien who had few friends among his co-workers. "They just didn't like him," Bert says. The men wouldn't pass food to him at the table, so Bert always saw to it he got enough to eat.

Schmidt evidently liked the finer things in life and strove to surround himself with what few luxuries he could in that rugged backwoods camp. Once when he was ill and not able to come to the cookhouse for his meals, Bert sent her flunky out to his cabin with dinner. The flunky returned wide-eyed to report that Schmidt wore silk pajamas!

During World War II Schmidt left camp to go east for a visit, traveling by rail. While changing trains in Nebraska, a zealous policeman overheard him talking in his broken English to the ticket agent and promptly arrested him as a German spy. It took Schmidt several hours before he could convince authorities he was not on the payroll of the German

Photo courtesy Oscar Aspli

Outside every cookhouse hung the iron triangle referred to as the "gut hammer" used to call the men to meals.

high command but was just a Tie Hack from the hills of Wyoming.

Bert's days as a cook were a dreary, round-the-clock routine of baking and cooking and keeping the fires going in her huge double-oven cookstoves, then falling into bed for some sleep before the day began again.

She could determine whether the ovens were hot enough to bake the never-ending stream of pastries and breads by "just putting my hand inside. If it felt warm enough I could start baking." She remembers proudly she never had a bread failure, baking at least three different kinds of bread a day.

That first winter, after she'd been snowed in for weeks and not even the supply sled had gotten through to camp, she was thoroughly disenchanted with her own baking. When the supply sled finally hove into sight, she welcomed the drivers profusely. They in turn had been smelling the delicious aroma of homemade bread baking as they approached the cookhouse, and their mouths were watering in anticipation. She gave them two loaves in return for something she had been longing for — a loaf of "storebought" bread.

Commenting on her cooking, Bert laughs, "I don't know how I got away with it; sure never learned to cook at home!" But she more than got away with it and soon was known for her melt-in-the-mouth cinnamon rolls.

At Thanksgiving and Christmas she would roast as many as ten turkeys at a time in her big ovens and turn out innumerable pumpkin and berry pies. "My tables looked beautiful, just beautiful," she says with a faraway look.

Her day began before 5 a.m., when she rose to get the fires started. Though she had a flunky to haul water and wash dishes and serve the food, Bert liked to make her own fires. She had a small heating stove near her room in the back of the cookhouse, and one morning when she was hurrying to get breakfast she noticed flames shooting up around the ceiling near the heating stove. About then the flunky

Teamster Bert Engstrom sits atop a load of supplies headed for the upper camps, some as far as twenty miles from Headquarters. The supplies were sledded to the cookhouses once a month. Orders for the families and the men who were batching went out at the same time.

arrived for work. Bert went right on with her meal preparations but said with a curt nod over her shoulder: "The roof's on fire up there."

"Well, that flunky just about jumped out of his hide! But he got busy and put the fire out."

At one point Bert was feeding so many men she needed an extra flunky. The company sent a young man with a terrible body odor. Bert took Jake, her regular flunky, aside and told him she couldn't stand that smell and couldn't he please do something?

Jake came up with what he was sure would clean up the matter. He suggested the two men take baths. "Jake told me he watched to see he took a good one, but he reported afterwards he smelled just as bad after the bath as he did before."

Fortunately the malodorous flunky soon tired of his new job and quit.

At another time Bert had an Irish flunky, Lee, who fell against the cookstove and burned his arm. The burn was beyond the simple emergency treatment Bert could provide from the Red Cross kit, so he was sent off to the lower country to see a doctor. In the meantime Bert badly needed help in the cookhouse.

Anxious to keep his good cook happy, the camp boss, John Selig, volunteered to wash dishes. A burly giant of a man, Selig performed his new duties docilely and without flinching. The other Tie Hacks watched stony-faced while John plunged his hairy arms into the soapsuds after each meal. It is reported that hoots of laughter could be heard after the crew left the cookhouse, and it's fairly certain the mirth was caused by the sight of their boss in this unaccustomed role.

At suppertime the white oilcloth-covered tables would be set with homemade bread, two kinds of meat, lots of potatoes and gravy, and vegetables. "Fruit soup" was a favorite at any meal, breakfast included. It was a concoction of dried fruits cooked and thickened with tapioca.

Courtesy Lydia Olson, circa 1925

The same kinds of food were prepared by the cooks for the river drive — only the method differed. All cooking was accomplished in dutch ovens over hot coals in open trenches. John Tangden stirs a pot of stew, watched by Adolph Solum and Alfred Bolin.

When the gut hammer (the big iron triangle hanging outside the door) was rung summoning the men to eat, they would come tromping over the rough, planked floors to sit on long benches on either side of the tables. Prodigious amounts of food were consumed in near silence and at amazing speed.

The first time Trego joined a crew at a cookhouse meal he sat down, leisurely helped himself to the food as it was passed, took a swallow of coffee, then picked up his fork to eat and stopped openmouthed. He was the only one eating! The others were already through.

There was an understanding among the men that there would be no conversation at the table. They were so given to arguing and fighting that they banished discussions at mealtime to avoid a possible fracas and the resultant wrath of the cook.

Some of the cooks were men, such as Fred Erickson, Andrew Berg, and Bud Warpness. While they didn't exactly serve French cuisine, their food had to be hearty and tasty or they had to answer to the hungry crews.

One cook turned out such poor fare the men threatened to kill him if he didn't improve. The next meal was no better, and the enraged crew took after him. The terrified cook fled into the woods, climbing a tree in an effort to elude his pursuers, who were hot on his heels. They held a gun on him until he promised never to cook another meal.

Pot-bellied Fred Erickson also cooked for the tie drives, moving the chuck wagon along with the drive and preparing

The cooks were expected to provide fresh breads and pastries daily for the men. Fred Erickson kneads a pan of bread dough after his pies for the day were baked.

Fred reigned supreme over this typical cookhouse. Each camp had a cookhouse and a camp cook. If more than fourteen men were being fed at a time, the cook was provided a flunky to help with dishes and meal preparation. The men ate at long, oilcloth-covered tables flanked with rough wooden benches.

the day's foods in the big dutch ovens that cooked over red-hot coals in open trenches. Bread had to be baked on the drives, too. One hot July day when we were visiting the drive my mother was watching Fred knead dough for the next meal. Flies were buzzing all around the open tent, and one flew into the dough. Fred just kept on kneading, casually flicking most of the insect out of his way with one finger.

"Raisins," he proclaimed, with a conspiratorial wink.

Next to whiskey, jovial Fred liked women best. There was a log cabin saloon about twenty miles from Headquarters called, by those who frequented the establishment, the Snake Farm.

The Snake Farm offered something or someone for every

taste. One time Fred left his cooking job to go on a stupendous toot. At the end of several days he was out of money and called Martin Olson to plead for more.

"Could I have a little more money?"

"No," thundered Martin, "You can't have any more money. You're in the hole."

Fred laughed blearily and declared happily, "Well, I'm in a hole out here, too, and it's a damned *good* hole."

Chapter Eight

THE WOMEN

There were few women in this male society. Though several of the Hacks may have had sweethearts or even wives in the Old Country, hardly any saved enough money to bring them to America.

Today the old-timers are gone, and few men remain who can recall what daily life was like in the camps. But several of the women who lived there still remember what it was to try to make a home in the near-wilderness, to give birth, and to raise families.

Though their lives were hard, to a woman those interviewed vowed the years on the mountain were "the best years of our lives."

It didn't always appear it would be, when they saw tie camp for the first time.

"I thought, 'This is the end of the world!' I never saw a telephone pole, I never saw a light pole, nothing," remembers Betty Dolenc, wife of the storekeeper.

The first day we moved to the mountain my mother, Agnes Trego, swore she would never leave the place if it meant she had to ride a car back down that narrow road. The day came, however, when she would drive it herself and not even panic when she met a hay truck.

When Dorothea Aspli arrived in the upper camps after a long, monotonous boat voyage from Norway and an equally long train trip across the continent from New York, she couldn't speak a word of English. With no time to become accustomed to this strange land, she began her new life as a

Photo courtesy Oscar Aspli

Dorothea Aspli, cook at Spring Creek, and her flunky Yvgve Hagstrom pose with the night's supper, big trout caught by Little John Olson at Moon Lake.

Photo courtesy Shirley Daniels

Shirley Daniels outside the honeymoon cottage Lewis built for her

cook at South Fork camp, with a dozen men to feed every day.

Shirley Daniels was nineteen when her new husband, Lewis, brought her to one of the outlying camps to live in 1936. With an outgoing, happy personality, she made an immediate hit with the Tie Hacks, who accepted the two young people as their own kind.

Their honeymoon cottage was a tiny board house with a slab roof built by Lewis himself. Other than the stove and bed, he made all the furniture, too. Since a camp would be abandoned when the timber was cut in that area, there was no incentive to build a structure which would outlive its usefulness. Shirley recalls at one camp they moved into an existing cabin that was "so small, I could lie in bed and cook breakfast."

That cabin was in bad repair, too. One day she was horseback riding when a sudden Wyoming downpour sent her galloping home. Returning to the little cabin soaked to

Photo courtesy Shirley Daniels

Lewis Daniels makes furniture for their first home

Photo courtesy Shirley Daniels

Corduroy roads they were called — logs and planks laid across low and swampy sections of the back roads. Martin Olson (*left*) watches Tony Dolenc changing tire on the "jitney." Lewis Daniels to the right.

Photo courtesy Shirley Daniels

Hanging out the wash. Even Tie Hacks had to attend to such mundane matters as clothes washing and housecleaning when they lived alone.

the skin and with her wet hair stringing around her face, she leaned out the window to wave to Lewis, who was passing by with a team of horses.

"My God, is that cabin leaking *that* bad?" he exclaimed.

Like many of the Tie Hacks' wives at one time or another, Shirley put in a stint of cooking in a cookhouse. One year she substituted for the schoolteacher at Headquarters, having been a teacher in a small rural community before she was married. And she took in laundry.

"I got tired of scrubbing clothes on a board, so I took in washing and ironing from the guys to pay for a gasoline-engine washing machine." She recalls ironing up to twenty-five shirts a week for the men, using old-fashioned irons heated on the cookstove, scrupulously darning their socks and mending their underwear. "They even brought me their suits to clean. I'd fill the old washer with naptha, throw in their suits, and turn it on. Can you imagine anything more dangerous, with that gasoline motor chugging away? Why,

Winter's end

any spark could have set it off. I guess it just wasn't my time to die."

Their first baby was due to be born the first week in December. When the time was imminent, Shirley left the mountain for Riverton, to be able to get to the hospital on time. By December 24 nothing had happened, and she insisted on going back to the tie camp for the Christmas dance. "I

Photo courtesy Oscar Aspli, circa 1930

Ready for the hunt. Big John Olson (*left*) and Jake Ottom set out to get their elk. Note the one long ski pole (*far left*) used by John for balance and as a brake when needed.

Photo courtesy Lydia Olson, circa 1914

Wyoming Tie & Timber Co.'s first Headquarters camp at Sheridan Creek. Lava Mountain is in the background. The site is now the Triangle C guest ranch.

Photo courtesy Lydia Olson, circa 1927

The original DuNoir, second Headquarters camp, nestled in the shadow of Ramshorn Peak. The wild beauty of the mountains was scarcely disturbed by the camps, which were always torn down and the area restored to its original state when there was no longer any use for them. Lydia Olson was cook at DuNoir in 1928, when the camp was abandoned. She was the last person to leave after cleaning up after the final meal in the cookhouse.

Courtesy Mike Guilford

Interior of the Rustic Pine Tavern in Dubois, one of the bars where the Tie Hacks spent their Saturday nights. Whiskey was twenty-five cents a shot, and every third drink was free.

danced all night, then drove back the hundred miles to Riverton. Ed was born December 29."

That was a year of long isolation in the camps. She and the new baby returned to a world of cold and snow that kept everybody snowbound for the next five months. The 2 a.m. feeding meant breaking ice in the water bucket to mix formula and building a fire in the cookstove to warm the milk and herself.

Yet she declares wistfully, "Those were the happiest days I ever spent. People weren't so greedy then. Friends were more important than things. None of us had any money. We made just enough to pay the commissary bill every month. Why, if we made $60.00 a month, that was a big amount when we were first married. But of course you could buy a month's groceries for $30.00. At the end of the month when everybody was low on supplies (they were only delivered

Martin and Lydia Olson. Photo taken about 1947

once a month) you'd go to the neighbors to see if they had any beans or anything left, and we'd all share." She lives alone in Eatonville, Washington, today, her husband dead for many years.

"We went to Dubois every Saturday night we could get out. Had to coast down all the hills 'cause we hardly had enough gas to make it. Then Lewis would shoot craps at the tavern to make enough money to get home on. He was always pretty lucky at that. A shot of whiskey was only twenty-five cents, and every third drink was free."

Lydia was married to the woods boss, Martin Olson, in 1928. By then she had been around the tie camps some twelve years, her first husband having been the forest ranger

When friends from out-of-state came to visit, a trip to Lake of the Woods for fishing was the order of the day. Trego drives the only vehicle capable of negotiating the mud of the backwoods roads in early June — or anytime after a hard rain. Behind are the visitors, daughter Joan (author) and wife Agnes.

Wyoming's rail fences can present a real challenge to the skier. Betty Dolenc laughs good-naturedly at her predicament. Agnes Trego in backgound.

assigned to supervise the tie-cutting operations of the fledgling Wyoming Tie and Timber Co. back in 1916. They lived at Sheridan Creek Ranger Station, next to the first Headquarters camp.

It was here the young woman gave birth to tiny, two-and-a-half-pound baby Doris. Only seventeen at the time, Lydia recalls, "I didn't expect her for another two months, so I didn't think I was giving birth to a child. But I had these awful pains that night! I was so young, I didn't know anything."

Alternately pacing the floor and trying to rest, she finally gave in to the terrible need to bear down while sitting on the slop jar. The baby was born — into the slop jar.

"We lifted her up on the bed with me, and Cay [her husband] woke up the young surveyor who had happened to come through the night before and was staying with us."

The surveyor dressed as fast as he could, rounded up horses from the pasture, and frantically rode off to get help. The nearest woman, Mrs. Moriarity, lived about two miles away. She hurried back with the flustered surveyor and quickly took charge. One look told her the premature baby didn't have a chance. Cutting the cord, she wrapped the infant and put her in the warmest place available — the still-warm oven of the cookstove — and turned all her attention to Lydia.

Spunky little Doris thrived in her homemade incubator. Looking back on it, Lydia is sure the only thing that saved Doris was the fact that the cord wasn't cut for so long after her birth.

"She was the cutest little baby you ever saw. She couldn't nurse, so we pumped my breasts and fed her through a medicine dropper. She was so tiny my bracelet went right over her head and my ring over her arm. Her little fingers were nearly transparent.

"Oh, I bathed her, and I fed her, and I played with her. I kept her by the cookstove, on a pillow on the chair, I did it — it was all in the day's work."

Two years later Cay died in the flu epidemic that swept the nation. With a youngster to support, Lydia went to work for Martin as a cook on Deception Creek. She married a second time, gave birth to a son, Norman, but five years later "it was over, we just couldn't make it." Again she returned to the tie camps. Martin had long had his eye on her, and this time he sought the help of another female in camp. "Are you gonna help me get that woman?" he asked. She apparently did, because Lydia and Martin were married in 1928 and moved to the third and final Headquarters camp as it was being completed. Lydia lives today in Dubois, near her daughter. In her eighties, she drives a jeep wagon that will take her back to the Warm Springs Mountain she loved so well.

The author, Joan Trego Pinkerton, age fifteen, sitting on corral gate at Headquarters

Chapter Nine

THE NIGHT THE MARTIANS LANDED

Back in 1938 a young actor named Orson Welles played a monumental trick on the nation with his radio broadcast, "War of the Worlds." It was so realistic thousands were convinced the Martians were at that very moment invading the eastern seaboard.

Lydia Olson was one of those listening to the radio who believed the unbelievable was happening. Her son Norman was in school 100 miles away, and every motherly instinct

Photo courtesy Lydia Olson

On a bright winter day the drivers line up their teams for an official photograph

The largest team was this beautiful matched pair of dappled greys, each weighing a ton. John Lund holds lead ropes.

The company's several teams of workhorses were as integral to tie cutting as were broadaxes and saws. The patient, plodding beasts skidded logs, pulled sleds and wagons. "They were as smart as we were," declares driver Bert Engstrom, who recalls more than one time being lost in a blizzard and having to rely upon his lead horse's ability to find the way back to camp.

told her to get to him so the family could face the invaders together. She convinced her husband, Martin, they must leave immediately. They were ready to set out when Trego knocked on their door with an even more ominous piece of news for Martin: Tony Dolenc was lost somewhere while elk hunting on the Continental Divide behind Headquarters.

Tony, the storekeeper, Trego, and a Hack named Vic Lawrence had been elk hunting that day when a snowstorm came up. Suddenly Trego and Vic became aware Tony was no longer with them. They fired their rifles, shouted and searched, but without success. With darkness setting in, they returned to Headquarters for help.

"We can't go now, we gotta find Tony first," Martin de-

It took a steadying hand at the head of the horse before Sorn Pederson, blacksmith, could nail on new shoes. Julius Vahrencamp *(left)*, Albert Edstrom *(center)*.

Supply sleds move upriver, drivers braced against the cold wind and snow

Lookout Mountain looms to the east of Headquarters, its name derived from the 100-foot fire lookout at the top. It was manned every fire season by employees of the Forest Service, who were skilled in fire-spotting. One of the worst fires, however, occurred before the summer even began in 1940. A sawdust pile, supposedly burned out the previous fall, smoldered all winter and burst into flames the first day of June. Flames raged over 2,000 acres before the fire was extinguished.

Sheets of flame sweep across the forest as a fire "crowns"

clared, leaving a near-hysterical Lydia and going with as many men as they could muster to begin the search.

While they were gone Lydia switched radio stations and learned the "invasion" had been a hoax. In the meantime, the search went on.

Tony himself tells the story of that night. He had sighted a herd of elk through the blowing snow and started after them, shooting one and dressing it out. "I built a little fire and waited for Trego and Vic to join me. I heard shots and figured they'd got their elk too. I answered those shots, but it was snowing — my gosh, was it snowing — and blowing. I looked around and knew I was lost. We'd crossed over the top of the Continental Divide, and it was my first time on the west side of the drainage. I just wasn't familiar with it.

"I found a sled track and followed it as far as I could.

Then I built a shelter and stayed all night in it. Next morning I started out again and found Lake of the Woods. Now I knew where I was. Then I walked out to one of the camps.

"There was little Ross Cobb — he was about four at the time — sitting on the step of their cabin. He looked up to me and said 'Why, God damn it, Tony, I thought you was supposed to be lost!' "

Since all the men in the camp were out looking for the lost Tony, it was left to Ross's mother to bundle the shivering man into a car and take him back to Headquarters.

In the daylight Martin led the searchers directly to where Tony had shot his elk. One of the men, Andrew Bloom, later brought the dressed-out animal into Headquarters for a grateful Tony, who had no great desire to return for the animal.

A horse grazes unconcernedly in rancher's pasture while smoke from forest fire cloaks the mountainside.

A stand of timber explodes into flames at height of forest fire

Andrew did it for his friend. He was so painfully shy he found it nearly impossible to talk to people unless he'd had a drink or so first. He'd bring in a list of the supplies he needed, hand it silently to Tony, and walk to the far end of the store while his order was being filled. But he evidently felt Tony was his friend. Once when he was drunk he went to Trego and had a $5.00 check written out to Tony's wife, Betty. He presented it to her with a flourish, saying, "Meesus, you gotta good husband."

Andrew lived alone in a crude bachelor cabin. When he needed to talk to Martin he would make his way to the Olson home, standing as far away from it as he could and still be heard. Half-hidden behind a tree he would cough loudly until someone heard him. Then Martin would be notified that Andrew wanted to talk to him.

One winter day Andrew had to break his self-imposed silence to come into Headquarters with the disturbing news that he had discovered the bachelor cabin of Hans Lundgren all frozen up and no sign of Hans anywhere.

Tony, Trego, and Bert Engstrom, a tie hauler, were the only men in camp at the time. Bert hitched a team to the small sled known as a go-devil, and they set out to search, fearing the worst. Andy had gone on ahead and had discovered an avalanche near Lundgren's cabin. He had begun digging a trench across the foot of the snowslide and had just come upon the frozen body of the unfortunate Hans when the three men arrived with the sled. Hans had been wearing bear-claw snowshoes instead of his usual skis and had been unable to get away from the tumbling snow as it swept down upon him. He had fallen forward on his stomach, and the snow pushed his legs backward up over his head.

The four men wrapped the doubled-up body in canvas and laid it between the two short seats on the go-devil. Since there was no place to sit but on the seats, Tony and Trego rode back into Headquarters with their feet propped atop the corpse, much to the disgust of their waiting wives.

Tony Dolenc holds grim evidence of cause of Hans's death: bear-claw snowshoes which prevented his escaping the onslaught of a snowslide (visible behind Tony). If Hans had been wearing skis he probably would have been able to outdistance the tumbling avalanche.

Hans's cap *(right)* and leg boot still attached to snowshoe were first uncovered by rescuers digging for the body. The picture was taken to document the death for the coroner.

Bert Engstrom drives the team pulling the "go-devil" which carried Hans's frozen body to Headquarters. On the sled (with feet propped on the canvas-wrapped corpse) are Andrew Bloom and Tony Dolenc.

Chapter Ten

WHAT A WAY TO SPEND A WAR!

It was 1944, and all available young men were off fighting for their country. Wyoming T&T faced a critical manpower shortage. The war was creating an ever-greater demand for railroad ties, but there were fewer and fewer men to produce them. The old-timers who were still hand-hewing had dwindled to a handful, and the sawyers were hard pressed to keep the mills supplied with trees. The company desperately needed tree cutters and "river pigs" to work the river drive.

In 1944-45 a German POW contingent lived in a little tent city hidden in trees many miles back in the woods.

The POWs even had electricity in their tents, which is more than the Tie Hacks had in their rough cabins. The POWs cut trees, worked in the sawmills.

Unexpectedly, help came in the form of the enemy. An entire German prisoner of war contingent was secured through VanMetre's efforts. The government deduced, wisely enough, that it was better for the Italian and German prisoners in their care to be doing something productive rather than simply languishing in prison camps. All over the nation, POWs were helping with farm harvests and other such tasks.

And so it was that a group of POWs and their U.S. Army guards were moved bag and baggage to one of the upper camps in 1944. The prisoners were willing workers in the woods. As one man recalled later, "What a way to spend a war!"

They were assigned to work in the outdoors at hard but agreeable work, given good food to eat, and a comfortable place to sleep. The tenthouses the POWs lived in may not

The compound was surrounded by a high wire fence and guarded by the U.S. military to discourage any would-be escape tries. However, the POWs found the backwoods life much to their liking — you couldn't pay them to escape.

Chow time for the POWs on the river drive. Food was served the way it had been for thirty years — from heavy dutch ovens strung out across the ground. Only the accent was different — no longer the singsong Swedish or Norwegian but now guttural German.

The 1945 river drive was manned mostly by German POWs, who proved adept at shepherding the ties down the swift water of the Wind River to Riverton.

have been the Ritz, but neither were they much different from what the Tie Hacks called home.

A high, wire fence surrounded the camp, which was guarded by only a handful of soldiers. The guards' job was equally amenable, since none of the prisoners showed the slightest inclination to leave.

Although women were not normally allowed in the POW enclosure, Betty Dolenc recalls that she visited with her storekeeper husband one day, taking along their infant son, Tommy.

The men poured out of their tents to make over the youngster, some of them unashamedly crying while they tried to explain with broken English and gestures that they had babies his age back in their homeland.

One of the prisoners who had been the personal barber of an army general in his native Austria begged to be allowed to cut Tommy's hair. "And he did a beautiful job, too," Betty remembers.

In the summer of 1945 the POWs were used on the tie drive and gave every indication they were having a ball while they were at it. The July Riverton newspaper account noted there were "fifty POWs, five Indians, and ten Swedes working the drive."

At one point Bert Engstrom, tie hauler, was assigned a POW to help him. Bert started to talk to him, but the man shook his head and gestured he couldn't speak or understand English.

As best he could, Bert showed him what he wanted

There were only five thousand hand-hewn ties, such as the ones the men are standing on, in this next-to-the-last river drive in 1945.

done. "Then that guy went ahead and did what he wanted — just the opposite way.

"So I sat down, took out a cigarette, and offered him one. He took it, and slowly he began to talk. He spoke better English than I did!"

Chapter Eleven

MECHANIZATION — AND AN END TO AN ERA

Mechanization and rising labor costs eventually changed the industry. Portable sawmills came into popular use in the early 1930s, and by 1936 many of the ties delivered to the yards each year had come from the mills.

In later years VanMetre recalled that back in the twenties

Top management meets with some of the men on a sunny winter day. From left to right: Martin Olson, superintendent of Wyoming Tie & Timber Co.; John Selig, camp boss and tie inspector; Jerry Cobb, sawyer; Alfred Olson, tie inspector and drive boss; Ricker VanMetre, president of the Wyoming Tie & Timber.

Engine for portable sawmill, on skids for reassembling at new cutting site

a man named Nick Carter had "suggested the three-man sawmills like they used in Vermont after we had shown him the old 'Clawhammer Mill' on West DuNoir Creek. The Clawhammer was steam-driven and meant winter sawing, where fourteen men regularly turned out 700 ties per day. There was no lost motion in that crew. Martin and I chuckled at the idea — but we came around to it when gasoline engines came into use."

These gasoline-driven mills were small enough to be run by two or three men and could be quickly erected or dismantled as need be. They were scattered throughout the Warm Springs drainage. Trees were cut and skidded by horse to the

By the mid-1930s the small portable gasoline sawmill had taken over the tie-cutting, and only a few of the skilled Hacks still plied their craft. The mills could be assembled in a day in a stand of timber, dismantled as quickly when the stand was cut. The rugged Wyoming scenery casts its tranquil spell over all the activities.

Close-up view of the simple cut-off saw cutting ties to regulation eight-foot length. Rough-cut lumber stacked in background was by-product of tie-making.

Sawyer Leo Armitage files saw blade

Logs are skidded to the mills by workhorses

Gasoline driven saws speeded up the tie-making operation, but the ties still had to be stacked by hand *(left rear)*.

mills. As they were brought in, the logs were sawed into ties and possibly some lumber if the tree was big enough.

When a stand of timber was exhausted, the mill could be taken down and moved that day to a new stand. More ties could be produced in a day than the most efficient Tie Hack could hew — he was becoming obsolete.

Improved roads and trucks gradually brought an end to the river drive, too. By 1943 ties were no longer being flumed. Instead, they were trucked down to the banks of the Wind River during the winter and spring and stacked in long sidings to be pushed into the water by bulldozer in the summer.

The final end of the tie drives came in 1946. Only 150,000 ties, all from sawmills, were yarded in Riverton that fall. The next year the Wyoming Tie and Timber Co. was sold to a Wisconsin businessman, J. N. Fisher, who purchased huge roller-bed trucks which hauled a carload of ties — 360 to 420 — at a time. Loaded trucks left the woods for Riverton daily.

Truck-trailer with roller bed that signalled final end of the tie drives. Trucks could more efficiently and cheaply transport ties from mills to yard.

Forklifts costing $15,000 each replaced the brawny Hacks who had hefted the ties to their shoulders and stacked them on the banks of the river. Even the wise old horses who had pulled the logs to the mill were replaced by bulldozers.

Most of all, the breed known as the Tie Hack was dying out. And management was aging. In 1947 VanMetre was sixty-five, Olson sixty-seven, and Billy Mac seventy-one.

Most of the original Tie Hacks were dead or gone — few had married to leave sons to follow in their footsteps. The Johnson Immigration Act of 1924 brought a gradual end to the flow of new recruits from the Old Country, and there were no new men to replace the ones who were gone.

In the 1945 drive there had been only 5,000 hand-hewn ties — cut by sixteen aging Tie Hacks who had been given a stand of timber on Snowshoe Creek to cut or not cut as they pleased. The camp was known as the Old Man's Home, and the Hacks cut mostly out of loyalty to Martin, who they knew would feed them and care for them whether they produced or not.

The era was over.

Martin said, "Maybe we should have educated a new generation of men for work in the woods. Anyway we came to an end.

"It was time to quit."

As a tribute to the men who created the legend, Wyoming T&T erected a fourteen-foot monument: The likeness of a Tie Hack carved out of a three-and-a-half-ton block of Bedford limestone. It depicts the Hack in bas relief with broadax and saw. In the background are scenes of the woods and river work in miniature — a skidder and team pulling a log, a river man with pike pole on the drive.

A bronze plaque reads simply:

> Erected to perpetuate the memory of the hardy woods and river men who made and delivered the crossties for the building and maintenance of the Chicago and Northwestern railway in this western country.
>
> Wyoming Tie and Timber Company
> 1946

William ("Billy Mac") McLaughlin, Martin Olson, and Ricker VanMetre personally supervise erection of the Tie Hack Memorial in 1947. The Bedford limestone carving stands at the grave of Bill Phillips, teamster killed while trying to apply a rough lock to his sled while hauling ties to the river in 1915. (Actually there were only two fatal accidents in the thirty-three years of the company's existence.)

VanMetre's son-in-law, Boris Gilbertson, sculpted the monument over a period of three years. It was erected on a knoll nineteen miles northwest of Dubois along U.S. Highway 26-287, overlooking the original Headquarters of what has been aptly called the greatest crosstie operation in the history of railroading.

At the erection were a handful of people, the Olsons, the VanMetres, the Dolencs, the Tregos, and Billy Mac — the surviving people to whom it meant the most.

In 1948 a dedication ceremony brought hundreds of people, including Forest Service and railroad personnel, to pay their tributes to the passing of a significant era in American history. A former forest supervisor on the Wind River, Roy Williams, wrote at that time:

Photo courtesy Riverton Ranger

A lone, hand-hewn crosstie standing in the Riverton Cemetery marks the burial plot owned by the Wyoming Tie & Timber Co. Some thirty old-timers are buried here.

Photo courtesy Riverton Ranger

Widows of three of the men involved in the management of the Wyoming T&T were on hand in 1974 for a special rededication of the Tie Hack Memorial after the U.S. Forest Service, Wyoming Recreation Commission, Wyoming Highway Department, Federal Highway Administration, and Bureau of Outdoor Recreation took over maintenance of the site and developed an information area adjacent to the monument for the benefit of tourists. The women are, left to right: Mrs. Ricker VanMetre, widow of the president; Mrs. A. B. Trego, widow of the secretary-bookkeeper; and Mrs. Martin Olson, widow of the woods boss, or Tie Pin, as he was called.

"I saw the transition from man and animal to man and machine, from winter to summer operation, and from sleds to trucks. I saw the last drive come down the flume and the last of the big river drives. I saw the passing of many Tie Hacks to a land which is bigger and better and where the drives never jam."

THE TIE HACK BOSS
DUNLOGGIN
SHOSHONE National Forest

BOOMS
THE TIE DRIVE

FLUMES

TIE HACK
MEMORIAL WAYSIDE

curious can pause to read about peaveys and pike poles and the woods boss. They can see an actual section of the old Warm Springs flume. All this information hints of a kind of life few today can comprehend.

The final chapter in the story of the Tie Camp was not to be written until 1953.

When Wyoming Tie and Timber Co. moved from Headquarters camp on Warm Springs Mountain in the fall of 1946 the houses and buildings were left vacant. The land belonged to the Forest Service, and nothing could be done with the buildings unless the use was approved by that arm of the federal government.

VanMetre worked doggedly over the next few years to

Photo courtesy Riverton Ranger

The Tie Hack monument was carved from a three-and-a-half-ton block of Bedford limestone and stands fourteen feet high. It was first erected in 1947.

A section of the original flume and descriptions of the Tie Hacks' tools and the jobs they performed give the tourist a taste of what life was like in the woods. The monument no longer stands in its original, stark simplicity — now it is surrounded by an enclosure made of railroad ties, and wide graveled paths lead up the knoll to the site.

Ricker VanMetre, president of the Wyoming Tie & Timber Co.

donate the buildings to a college or church for summer camps — a permissible use.

An investment of perhaps $10,000 would have been required on the part of any new owner to repair leaky roofs and other weather damage, as well as to install new sanitary facilities to bring the buildings into compliance with Forest Service standards.

For a time a Baptist church expressd enthusiasm for the project, then faded from the picture. In January 1949 Yale University appeared excited about setting up a permanent

For one week every summer in the early 1940s the old bunkhouses and cookhouses were used by Wyoming 4-H youth for a summer camp. Here kids wait cup and plate in hand, for supper. Bunkhouses are in background.

The original homestead cabin at Headquarters could still be seen in the late 1950s

geology summer camp. During correspondence with the head of the geology department VanMetre noted it was his friendship with a Yale man that "prompted the idea of offering the place to Yale, although I am a Princeton man myself."

In June the university decided it could not afford the project. VanMetre wrote to the Forest Service, entreating it to help in the financing:

"It has long been my dream and hope," he wrote, "that the old camp could be put to such use, in the firm belief that the explorations by such a group over the future years very conceivably could discover mineral or oil deposits of great value, and in general be a benefit to the local community.

"I hope you can envision the potential benefits of such a project to the Forest Service and public welfare as I do. It seems to me that cooperation and financial assistance of the Forest Service could be justified as a matter of Forest Resources Development. The amount of money needed is very small in comparison with the possibilities of such a venture."

Two weeks later he had his reply. The Forest Service said no. VanMetre acknowledged the refusal in a poignant letter:

"This seems to be the end of my dream for putting the old camp to permanent useful service, and the time has come for us to turn it back to the Forest Service for such use as can be made of the structures for the public benefit.

"The reports I receive of abuse of the buildings and vandalism by the public do not increase my respect for the rabble that infests the place, but I hope the Forest Service can carry out a plan for recreational use that will work out satisfactorily."

After Headquarters was abandoned, cabins began to deteriorate before a use could be found for the old camp.

He thanked the foresters for their cooperation and signed the letter. Then he made one last attempt:

"P.S. Maybe some public works projects are in the not-distant picture that would fit in with a program of rehabilitation for the camp for public use?"

VanMetre sent the Olsons a copy of the letter, adding this sad note across the bottom in his handsome scrawl: "My last shot missed the mark. Forest Service lacks the money to fix up the place for the Geology Department."

The special use permit under which the Wyoming T&T operated all its camps specified that when the use designated was terminated, the land must be returned to its original condition. In the next few years some of the houses were sold to private purchasers and moved off the mountain. Everything else was razed and burned, the grounds cleaned up, and the sagebrush raked.

A final letter March 10, 1953, from Forest Supervisor E. L. Miller to VanMetre reads:

"We are informed that the clean-up work has been completed at the old headquarters camp at DuNoir and that all the buildings have been removed. This completes the work for which your company was responsible, and the case is being closed on our records. In a way, we were sorry to see the old camp disappear, as it had been a landmark for a good many years; however, since the camp has been abandoned there had been so much vandalism that it was rapidly going to pieces, so it is probably best that it is gone."

Today only a few paths and rutted dirt roads indicate there was ever any activity. I stand on that broad plateau with the wind whipping around me and see in my mind's eye the rows of cabins and the hard-working families who lived there and whose children skied to a little, one-room schoolhouse up near the water spring. Over there was the rambling log store, its big pot-bellied stove fed by four-foot logs, warming the Tie Hacks who came to buy their boots and snoose.

I can hear the ring of the anvil as Sorn pounds out

another shoe for the great, gentle workhorses that graze about the houses, sometimes coming up to sniff at our pockets hoping for a sugar cube.

Faces of Hacks long dead drift through my mind, and I hear again their rhythmic singsong talk as the axes bite sharp into the trunk of a tree.

One thing endures: I gaze at the magnificent pine forests all about me, so carefully and selectively cut that they remain for future generations.

The forests and the Tie Hack Memorial — fitting tributes to a way of life that no longer exists.

Back in the woods, vines and grass gradually begin to take over an abandoned cabin

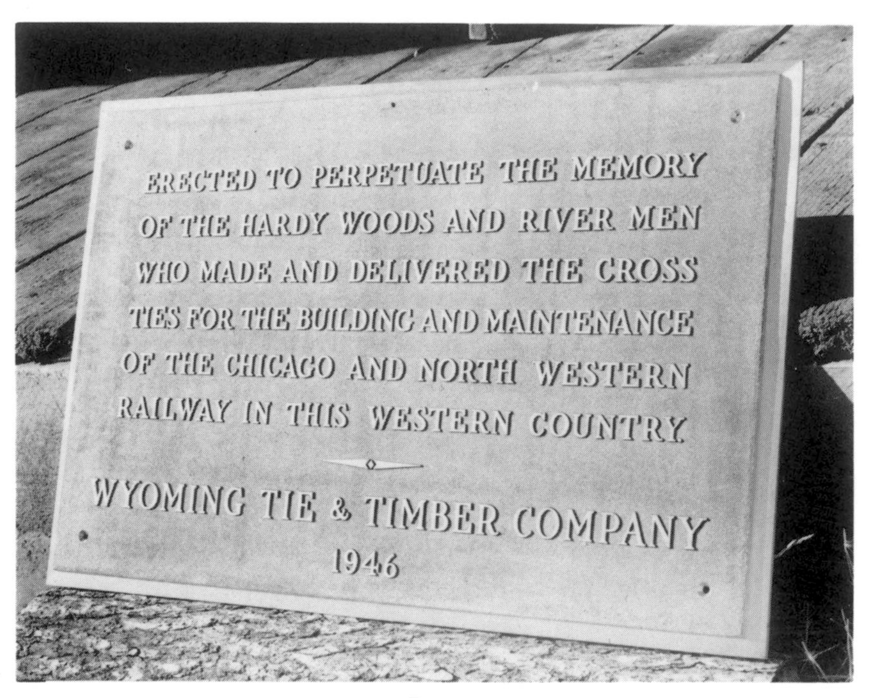
ERECTED TO PERPETUATE THE MEMORY
OF THE HARDY WOODS AND RIVER MEN
WHO MADE AND DELIVERED THE CROSS
TIES FOR THE BUILDING AND MAINTENANCE
OF THE CHICAGO AND NORTH WESTERN
RAILWAY IN THIS WESTERN COUNTRY.
WYOMING TIE & TIMBER COMPANY
1946

Finis